PHASE TRANSITIONS IN FERROELASTIC AND CO-ELASTIC CRYSTALS

An introduction for mineralogists, material scientists and physicists

Student edition

Ekhard K.H. Salje

Department of Earth Sciences, Cambridge University and IRC in Superconductivity

CAMBRIDGE UNIVERSITY PRESS

CAMBRIDGE UNIVERSITY PRESS
Cambridge, New York, Melbourne, Madrid, Cape Town, Singapore,
São Paulo, Delhi, Dubai, Tokyo, Mexico City

Cambridge University Press
The Edinburgh Building, Cambridge CB2 8RU, UK

Published in the United States of America by
Cambridge University Press, New York

www.cambridge.org
Information on this title: www.cambridge.org/9780521429368

© Cambridge University Press 1990

First published 1990
This student edition first published 1993

A catalogue record for this publication is available from the British Library

Library of Congress cataloguing in publication data available

ISBN 978-0-521-38449-0 Hardback
ISBN 978-0-521-42936-8 Paperback

To Lisa, Henrik, Joëlle, Jeanne, Léa-Cécile and Barbara

CONTENTS

Errata

E. K. H. Salje, *Phase transitions in ferroelastic and co-elastic crystals*
ISBN 0 521 42936 6

p105, equation 8.45, *for* $\coth 2\left(\dfrac{x}{w}\right)$ *read* $\coth^2\left(\dfrac{x}{w}\right)$

p133, equation 9.41, first line is missing, equation should read:

$$\langle \Delta Q \rangle \approx n\frac{4\pi}{v}\int_d^\infty |\bar{Q}-Q|\frac{1}{8\pi g}\frac{1}{r}e^{-\frac{d-r}{r_c}}r^2\,dr$$

$$= n\frac{4\pi}{V}|\bar{Q}-Q|\frac{1}{8\pi g}r_c^2\left(1+\frac{d}{r_c}\right) \qquad\qquad 9.41$$

p141, equation 9.51, last term of last line:
$$\text{for } \lambda QQ \text{ read } \lambda QQ_{od}$$

p166, equation 11.2, first term in each line
$$\text{for } \neq \text{ read } \dagger$$

p207, equation 12.6, *for* p *read* π

PREFACE

Research on ferroelastic crystals and related materials has obtained some aspects of a tradition but has not yet mellowed. Vigorous discussion on serious questions such as the underlying physical principles which generate elastic instabilities have just begun whilst debates concerning nomenclature still linger on. Application of the physical concepts in Earth Sciences, Material Sciences and Solid State Chemistry are new and growing rapidly. To write a monograph at this stage of development must be premature but if this book is perceived by the reader as a stimulus rather than the outline of the final concepts, it has fulfilled my hopes.

A few words of apology are called for if the reader is not to criticise the book more sharply than it doubtless deserves.

There are reviews on ferroelasticity, such as that by Wadhawan, and many an article dedicated to the illumination of symmetry concepts as started so early by Aizu, the Czech school of Janovec, Dvorak and Petzelt, the innovative French Toledano brothers, and the recent contributions by Hatch and Stokes. All this work deserves the highest credit. None of them, so far as I know, has quite the same purpose as this book, however. Pure symmetry considerations are aesthetically satisfying and are very useful, in particular for classification purposes, when all the physical facts are known. The concept that all ferroelastic phase transitions always follow the simple track of direct group-theoretical predictions of the most simplistic manner, however, died some years ago. Not that such phase transitions do not exist, it is simply that our imagination wandered off to more interesting questions concerning fundamental physical notions. I have thus reduced the aspects of group theory in this book to a bare minimum which should suffice for the understanding of my arguments. Tables are provided so that the reader can work out what he or she wants to know but cannot find here or in any of the quoted references.

The problem of selection, in a book of this genre, is very difficult. Examples had to be given to illustrate the ideas. Too many examples and too many ideas, however, make a book intolerably lengthy and may discourage the reader. The formation of anti-phase boundaries, although closely connected to the theory of twinning, has, as an example, not been discussed in this book. To ease the pressure on the missing completeness of a review, I have compiled an extensive list of ferroelastic and co-elastic compounds. Readers might find it convenient to use the tables in a lexicographical manner if they want to find stimulation for their own work.

Selection of ideas is even more difficult than that of examples. The most novel ideas always capture our imagination most, so that my own interest in this matter increased with the chapter number. It took, indeed, a page limit set by the publisher, a time limit set by my teaching duties, and strong words from my secretary, Alex Brewer, to stop these efforts where they are. I left it there with the clear feeling that not half of what should have been said has been said. The scholar who deplores that some lines of argument have only been mentioned but not worked out in any detail has my unreserved sympathy.

This book was written in the evenings and at some weekends in late 1989. This time was taken away from my family and I am grateful to them for not complaining with greater vigour than they did. It is a great pleasure to thank my colleagues and co-workers in Cambridge: V. Heine, A. Putnis, M.A. Carpenter, M.T. Dove, B. Wruck, B. Güttler, W. Schmahl, S.A.T. Redfern, D. Palmer, B.Winkler and M. Harris for pleasant and stimulating discussions. BW, MTD, and W. Selke in Julich, took the burden of proof-reading the manuscript. Sue Jackson's editorial expertise was invaluable. The work would not have been undertaken without the tremendous help of Alex Brewer.

The student edition is part of the first edition and has been corrected for mistakes. I am grateful to all colleagues who pointed out such mistakes, in particular to Professor Frey, Munich.

EKHS   Cambridge, 1992

This book is the student edition of a larger volume which was published by Cambridge University Press in 1990. It contains all those chapters that the author feels would be useful for undergraduate and graduate teaching. Chapters 1–6 and the introductory part of Chapter 7 are suitable for teaching a second and third-year course, such as the Part I B and Part II Natural Sciences Tripos in Cambridge. These chapters contain a compilation of empirical observations and their phenomenological description. Some of the simpler equations have not been derived in detail because it is assumed that either students will have some elementary knowledge of solid state physics sufficient for their own studies or that the lecturer and the supervisors will elaborate on these points during the oral presentation of the course.

Building up from here, sections 5.2, 7.1, 7.2a, 7.3a, 7.4a, 8.1, 8.4 and 9 should be within the reach of undergraduate students in their final year or at the beginning of a graduate course. The remaining chapters can conveniently be used for graduate teaching, seminars etc. They should lead the reader to further research work and might also be useful as reference material, particularly the bibliographies. The index refers mainly to specific definitions and should be used in conjunction with the contents, which refers to general topics.

As these various chapters are built into each other, following lines of argument rather than being self-contained lecture courses, some readers might be discouraged if they run prematurely into problems which they cannot yet solve. The concept of the book is such that these obstacles can be ignored in the first run without losing too much of the context when starting with the next chapter. It is then useful to reread the difficult chapters at a later stage. Finally, having done so, the reader should be at a level which enables him to follow the current research literature.

1

INTRODUCTION AND SOME DEFINITIONS

In the first chapter the basic phenomenological features of ferroelastic and co-elastic crystals are summarised: i.e. twin structures, lattice distortions and the ferroelastic hysteresis. As some of the basic definitions are introduced here, even the knowledgeable reader should scan through this chapter - he or she might also enjoy the figures.

Many crystals, when observed under the optical microscope, display characteristic domain patterns. One of the most common microstructures is related to twinning, with dominant twin planes oriented approximately perpendicular to each other. Such twins have various origins and may arise from growth phenomena or transformation twinning. These twin structures are often seen to change when the crystals are squeezed or even only lightly touched with a preparation needle. This phenomena has, in particular, been the bane of structural crystallographers who found that single crystals spontaneously twinned during preparation for subsequent X-ray or neutron experiments. Mechanical twinning is not only observed in technologically important materials such as high T_c superconductors but is also commonly encountered in mineralogically and petrologically relevant systems, such as perovskites, boracites, palmierites, leucites, quartz and feldspars, to name but a few. Figure 1.1 shows a rather typical domain pattern in a crystal with a palmierite structure ($Pb_3(PO_4)_2$) which was induced by a gentle squeezing between the finger tips.

Observations of such effects have long been noted in the mineralogical community and their importance has certainly been recognized by eminent mineralogists, such as Laves (1969) who called these twins 'Druckzwillinge' (i.e. stress-twins). Even earlier, Wooster and Wooster (1946) tried to make use of this effect to de-twin Dauphine twins in quartz by applying thermal gradients, twisting forces and the like. Metallurgists also observed mechanical domain patterns in alloys during this time and the term 'rubber-like' has been used to describe the behaviour of Au-Cd alloys. Some other alloys, such as fcc indium 23 at% thallium (Saunders et al. 1986), show a thermodynamic

Salje

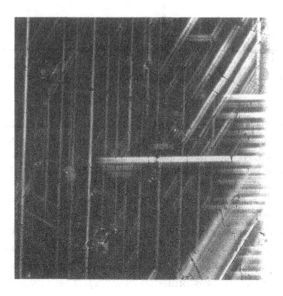

Fig.1.1 Typical domain structure in lead phosphate ($Pb_3(PO_4)_2$). Ferroelastic domains perpendicular and inclined to the crystal surface can be seen.

behaviour close to that observed in the above-mentioned materials and it appears that most of their physical properties can be described along the same lines as the crystals which we shall consider in this book (Dmitriev et al. 1989).

As well as optical microscopic observation, mechanical twinning can be investigated on a macroscopic scale. Let us consider the classic 'hysteresis' experiment as demonstrated by Salje and Hoppmann (1976) (Fig.1.2). Uniaxial stress is applied to a crystal using a weight. It will be shown later that the relevant stresses are usually shear stresses so that in this experiment the crystal is attached by one face to a solid surface while the opposite face is sheared by the attached weight. Simultaneously, the change of shape of the crystal is measured. It is necessary to obtain accurate measurements of the shear angle because the macroscopic shear of the crystal is normally small (i.e. several seconds of a degree). In this experiment, an optical interferrometer with a high angular resolution was used. It is also instructive to measure the related change of the birefringence of the crystal. Salje and Hoppmann used an arrangement of the optical polariser and analyser so that one type of twin domain was in the extinction position. The second type of twin domain then contributed towards the total optical birefringence. When the shear stress was applied, the proportion of the two

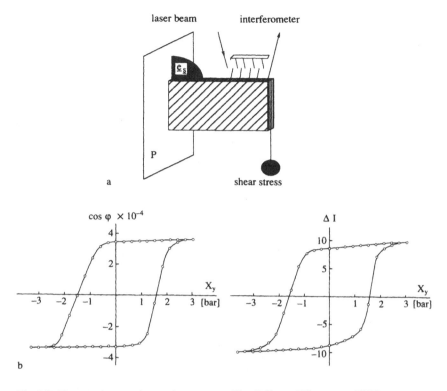

Fig.1.2 Hysteresis experiment demonstrated by Salje and Hoppman 1976:
a) The experimental arrangement for measurement of a mechanical hysteresis. The crystal is glued to an immobile plate P and sheared by a weight on the string attached to the opposite surface of the slab. The shear angle φ is measured using an optical interferometer. The shear angle is proportional to the macroscopic strain e_s which is determined as a function of the weight (i.e. the shear stress).
b) Hysteresis curves determined using the arrangement shown in Fig.1.2a. The strain e_s is plotted as cos φ, the stress is $X_y = P_6$ in units of bars. The hysteresis curve on the right hand side shows the optical birefringence in one domain as a function of the external stress as observed between cross polarizers. Both curves intersect the stress axis at 1.6 bars for zero strain. Their shapes differ due to non-linearities in the relationship between the spontaneous strain and the induced optical birefringence.

twin domains changed and the total birefringence changed as a function of the applied stress. The process of changing the orientation of twin domains due to external stress is often referred to as 'ferroelastic switching'.

The experimental results for a crystal of lead phosphate are shown in Fig.1.3. There are two shapes of crystal which occur if no stress is applied.

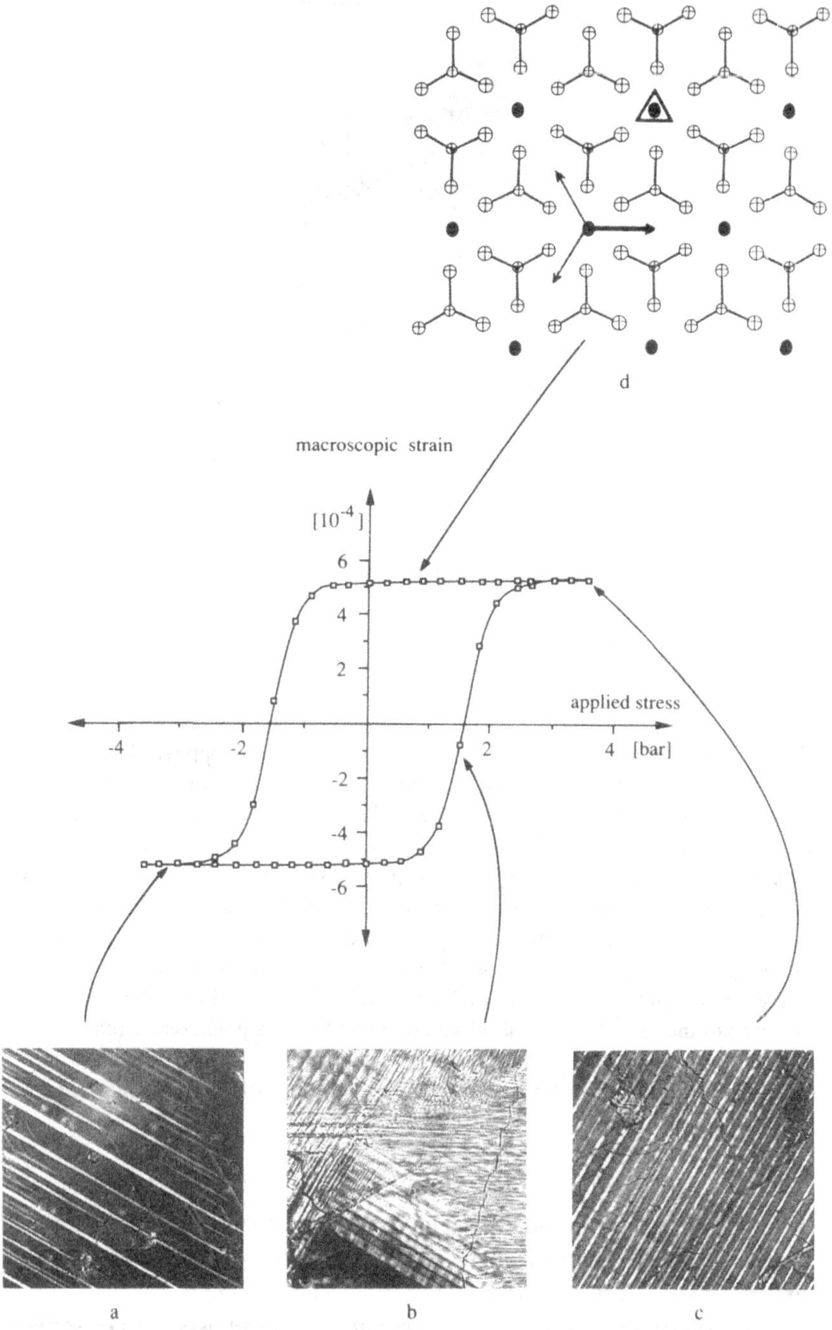
d
macroscopic strain
[10^-4]
6
4
2
-2
-4
-6
applied stress
-4
-2
2
4
[bar]
a
b
c

The strain associated with each of them (i.e. the value on the strain axis for zero stress) is called the **macroscopic spontaneous strain,** i.e. the deformation of the crystal with respect to a hypothetical structural state which is represented by the origin and would not allow for changing domain patterns. The zero point of the plot in Fig.1.3 is chosen in such a way that the two strain values are symmetrical with respect to the origin. When a stress is applied, we see in Fig.1.3 that one structural state remains stable and displays the classic elastic behaviour whereas the second state is unstable and collapses into the first if a critical stress is surpassed. If the direction of the stress is reversed (i.e. its sign is changed) we find that the second structural state is stable and the first collapses into the second. The strain-stress relationship in Fig.1.3 is called an **elastic hysteresis,** and a crystal showing an elastic hysteresis is called **ferroelastic.** A quantity which characterizes the size of the elastic hysteresis is that stress which is necessary to destroy the domain pattern. This stress is defined by the intersection of the extrapolated stability curve and the transformation branch (Fig.1.4) and is called the **critical stress.** It is roughly equal to the stress needed to induce significant changes of the domain pattern during observation by optical microscope. A second important stress parameter is the **coercive stress** which is defined as the intersection of the elastic hysteresis with the stress axis (i.e. at zero strain). Experimental determination of the coercive stress requires measurement of the full hysteresis curve and is hence much more involved than the determination of the critical stress. The most easily accessible experimental quantity is the equivalent optical hysteresis as shown in Fig.1.2. By comparing the mechanical and the optical hysteresis in Fig.1.2, it is clear that both curves reflect the same domain switching process. A more quantitative examination shows, however, that the macroscopic strain of the sample and its integral birefringence are not strictly proportional to each other. Deviations from proportionality occur, in particular, at stress values near the critical

Fig.1.3 Experimentally observed microstructures, hysteresis curve and crystal structure of lead phosphate. The two fully 'switched' crystals (**a**) and (**c**) display striped twin patterns whereas the crystal in the intermediate state (**b**) shows superpositions of various twin orientations. The equivalent atomic processes which occur during the switching are shown in (**d**). The crystal structure of ferroelastic $Pb_3(PO_4)_2$ consists of PO_4 tetrahedra which are arranged around a pseudo triad indicated by the triangle. All lead atoms (black) are displaced from the triad by some 0.07Å along the thick arrow. There are three orientations, however, shown by the arrows along which the lead atoms can displace. Ferroelastic switching occurs when the direction of the displacement is changed by external shear stress.

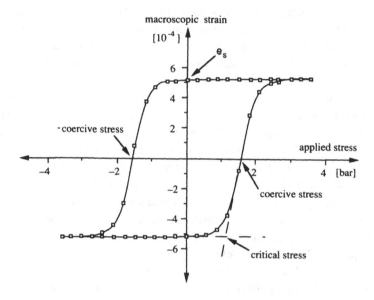

Fig.1.4 Mechanical hysteresis of $Pb_3(P_{0.8}V_{0.2}O_4)_2$ showing the characteristic quantities e_s indicating the macroscopic spontaneous distortion of the crystal without external stress. The critical stress is defined by the stress at the intersection of the two branches of the hysteresis curve. The coercive stress is defined by the intersection of the hysteresis with the stress axis at zero strain.

stress where the optical hysteresis often has a smoother appearance than the mechanical equivalent. The reason for this behaviour is that the optical birefringence does not only reflect the volume proportion of domains switched into the energetically more favourable orientation, but also the structural intra-domain distortions and the interaction between different domain walls. We shall see later how these effects can be quantified; here it is only important to note that the induced birefringence and the strain are highly correlated and that in many minerals, at least, the spontaneous strain and the ferroelastic birefringence (at zero stress) are proportional to each other (top and bottom curves in Fig.1.5). This proportionality (if valid) has been the key to a number of successful studies of ferroelastic materials (e.g. review by Glazer 1989).

The ferroelastic properties of a mineral invariably depend on the thermodynamic circumstances under which they are measured. In most materials we find that the size of the hysteresis decreases with increasing temperature and decreasing pressure. Very little experimental work has been done on ferroelastic behaviour under hydrostatic pressure and we shall

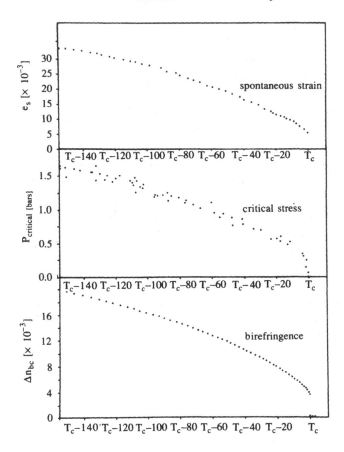

Fig.1.5 Temperature evolution of the spontaneous strain, the critical stress and the morphic birefringence in lead phosphate. All temperature dependencies are identical within experimental error, which indicates that the relative shapes of the hystereses curves in Fig.1.2 are unchanged whereas the total size decreases with increasing temperature. At temperatures above the transition point all parameters are strictly zero and the crystal is paraelastic.

concentrate here entirely on the effects of temperature and, for some examples, variations of the chemical composition. Let us first inspect the effect of temperature on the ferroelastic hysteresis in Fig.1.3. The temperature evolution of the three characteristic quantities, namely the spontaneous strain, the critical stress and the ferroelastic birefringence is shown in Fig.1.5. The comparison of all three curves shows that the three quantities are proportional to one another within the experimental resolution. This means that the shape of the ferroelastic hysteresis is always the same and that only its amplitude decreases as the temperature increases. Above a well defineu temperature, the ferroelastic effect disappears; the crystal ceases to be ferroelastic and is now called **paraelastic**. The transition between the paraelastic phase and the ferroelastic phase is called a **ferroelastic phase transition.** The ferroelastic quantities, spontaneous strain and coercive stress, disappear in the paraelastic phase by definition. This is not necessarily true for other related physical properties such as the birefringence. The ferroelastic effect will change the birefringence of the crystal in all directions. In some cases, a high symmetry orientation for the propagating light can be found, in which no birefringence exists in the paraelastic phase and where an optical birefringence due to the ferroelastic effect occurs in the ferroelastic phase. This special case of a ferroelastic birefringence is then called the **morphic birefringence**. In our example the optical birefringence was measured along the triad in a rhombohedral crystal and has hence to disappear in the paraelastic phase. The rhombohedral symmetry is destroyed during the ferroelastic phase transition so that the triad no longer exists in the ferroelastic phase and hence a morphic birefringence appears for light travelling along the same direction as the triad had been in the paraelastic phase. We shall find further examples of morphic and non-morphic birefringence in ferroelastic materials in later chapters.

2

FERROELASTIC AND CO-ELASTIC PHASE TRANSITIONS

Here the term 'co-elastic' is formally defined as a generalisation of
ferroelasticity. It is argued that both ferroelastic and co-elastic phase transitions
can be treated within the framework of Landau-Ginzburg theory and that is
exactly what will be done later.

When a ferroelastic crystal is heated we usually find that the ferroelastic
effect disappears at a well defined temperature. At this temperature a
structural phase transition between a ferroelastic and a paraelastic phase takes
place with the main feature being that a ferroelastic hysterisis exists in one
phase but not the other. A possible transition mechanism is that the thermal
excitation of domain switching occurs in such a way that the crystal maintains
the crystal structure of the ferroelastic phase on a local scale even in the
paraelastic phase. The domain boundaries between the different ferroelastic
domains would be highly mobile and thus no macroscopic domain switching
is possible; the crystal is macroscopically paraelastic. This transition
behaviour has indeed been observed in very few ferroelastic materials, such
as high purity lead phosphate, and we shall comment on the significance of
mobile domain boundaries in Chapter 8. In the vast majority of all ferroelastic
crystals, however, one finds that the physical mechanism of the transition
between the paraelastic and ferroelastic phase is only indirectly correlated
with the ferroelastic effect itself. Looking at it in a different way we may ask
ourselves what makes a crystal ferroelastic. We shall see later that, in fact, we
need two ingredients: firstly a phase transition between the paraelastic and the
ferroelastic phase which creates a lattice distortion. Secondly, that this lattice
distortion can be reorientated by external stress. It is useful to distinguish
between these two features because they do not necessarily reflect the same
physical processes. It appears that the role of the structural phase transition is
only to provide a lattice distortion which can build up to the macroscopic
spontaneous strain which then, in turn, can be switched by the coercive stress.
Although this lattice distortion can be created by a multitude of different
mechanisms there are common features in all ferroelastic phase transitions.
Before we explore these features we may ask whether it is useful to restrict

our discussion of the related phase transitions to those materials which are ferroelastic. Imagine that a crystal possesses a macroscopic spontaneous strain but all attempts to switch the domain structure lead to destruction of the sample. This crystal is not ferroelastic in the strict sense of the definition above, although most of its structural and physical properties will be similar to those of ferroelastics. Some authors have called such materials **'hard ferroelastics'** and all crystals which have the physical potential to be ferroelastic but where no actual hysteresis measurement has been successfully performed, have sometimes been named **'potential ferroelastics'**.

Other classes of materials which are somewhat similar to 'ferroelastics' consist of those crystals possessing a macroscopic spontaneous strain (the prerequisite of the phase transition we wish to consider) but in which there is no possible physical mechanism which could lead to its reorientation. Such behaviour always occurs if the size of the crystal is too small to form domain structures. These crystals are called 'nanocrystals' because their typical size is measured in units of nanometers rather than microns or larger as are 'normal' ferroelastics. Another class of material which cannot display ferroelastic hysteresis is that in which there are no symmetry equivalent ferroelastic states, despite the formation of a large spontaneous strain in the low symmetry phase. A typical example is calcite. Calcite undergoes a structural phase transition near 1250K between a high temperature phase with disordered CO_3 molecular groups and a low temperature phase in which the degree of orientational order increases gradually with decreasing temperature (Fig.2.1). A simple description of this phase transition in terms of orientational correlations between the CO_3 groups is not adequate, however, because the increasing degree of order with decreasing temperature leads **automatically** to a substantial decrease of the c-lattice parameters and to a lesser extent to variations of the a-lattice parameters (Dove and Powell 1989). This mechanism is also observed in the well studied $NaNO_3$ (Reeder et al. 1988, Redfern et al. 1989, Schmahl and Salje 1989). The variation of the c-lattice parameter leads to a large macroscopic spontaneous strain, in the same way as in ferroelastics. As this strain is aligned along the unique triad of this crystal structure it is impossible, for symmetry reasons, to switch the lattice distortion into any other equivalent orientation because there is none. The crystal is not ferroelastic, therefore, and not even 'potentially ferroelastic'. The phase transition mechanism, on the other hand, contains as an essential ingredient, the creation of the macroscopic spontaneous strain which stabilises the low temperature structure of calcite (e.g. Lynden-Bell et al. 1989). The build-up of the spontaneous strain and the stabilizing effect of the associated strain energy are features very similar to those observed in ferroelastics and it is often useful to discuss the features of the phase

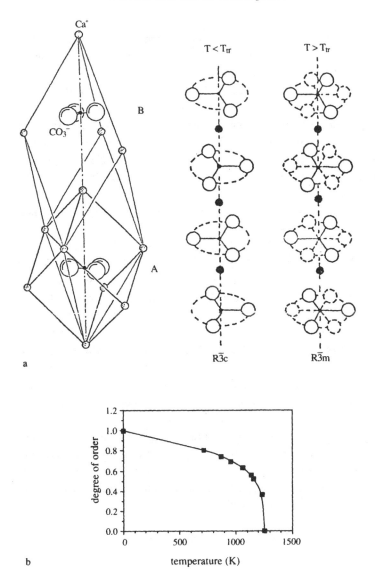

Fig.2.1 Structural phase transition in calcite:

a) Crystal structure of calcite ($CaCO_3$). The triangular CO_3 groups are ordered antiparallel to each other at T< T_c~1250K (large unit cell B, space group $R\bar{3}c$). At T >T_c, no long range order exists and the small unit cell A relates to the space group $R\bar{3}m$.

b) Degree of orientational long range order of CO_3 groups in $CaCO_3$ (calcite) as a function of temperature. The data points are related to intensity measurements of ($11\bar{2}3$) reflections by Dove and Powell (1989).

transitions of both groups of materials within the same context. It is clear that although ferroelastics can be defined easily from their physical properties, a similar simple definition does not exist for the second class of materials. Their general characteristics are quantitative rather than qualitative, namely the appearance of large spontaneous strains and large elastic energies leading to the structural phase transition. Many names have been used in literature in order to classify these materials including 'mechanically soft', 'hyperplastic', 'pyroelastic' and 'co-elastic'. We shall use the term **co-elastic** to imply that elastic and strain anomalies in these materials are correlated with the structural phase transition.

The common features of ferroelastic and co-elastic phase transitions are, therefore, their association with elastic instabilities, which have been analysed in great detail by Cowley (1976) and Folk et al. (1979). These authors have shown that if the transition mechanism involves long ranging acoustic waves these waves are strongly confined in their propagation direction, as we shall see in more detail in Chapters 7 and 9. In low symmetry materials, the relevant acoustic waves occur only in one dimension, and in some cubic, hexagonal and tetragonal systems they are restricted to two-dimensional sectors of the crystal (e.g. Knorr et al. 1986, Mayer and Cowley 1988). Inhomogeneous fluctuations of the lattice distortions due to the structural phase transition are then confined to the same subspaces, i.e. they occur either as one-dimensional or two-dimensional fluctuations. Renormalization theory shows, however, that such fluctuations are 'harmless' and do not lead to criticalities which could not be approximated by Landau theory (including possibly some logarithmic corrections). It appears safe to assume, therefore, that Landau theory will accurately describe the main physical features of ferroelastic and co-elastic phase transitions. Possible deviations from the predictions of Landau theory will be discussed in Chapter 8.

3

THE LANDAU POTENTIAL

Some basic ideas about Landau theory are introduced. Some readers might know all about it - but for those who don't the simple description given in this chapter suffices for the understanding of the rest of the book.

Let us consider a phase transition between two phases with different symmetries. We require that all symmetry elements of the low-symmetry phase are already present in the high-symmetry phase. In other words, the low-symmetry phase is characterized by a 'broken symmetry' of the high-symmetry phase. This requires that an additional thermodynamic variable exists which specifies the thermodynamic state of the low-symmetry phase. In a ferroelastic material, for example, the spontaneous strain disappears in the paraelastic phase because it is not compatible with the symmetry of this phase (e.g. the lattice deformation is dynamic rather than static in the high-symmetry phase).

Thus by breaking the symmetry a new variable is created which is called the order parameter Q. This variable is necessary to specify the thermodynamic state completely. Under certain thermodynamic conditions, the low-symmetry phase will be thermodynamically more stable than the high-symmetry phase. The difference in Gibbs energy between both phases, which stabilizes the low-symmetry phase, is called the **excess Gibbs free energy** G, which has now to depend on the thermodynamic parameters temperature T, pressure P, chemical composition N *and* the order parameter Q:

$$G = G \ (T,P,N,Q) \qquad\qquad 3.1$$

We can then calculate the Gibbs free energy $G(T,P,N)$ that we observe in an experiment under equilibrium conditions, by using the minimum principle:

$$\partial G/\partial Q = 0 \qquad\qquad 3.2$$

In what follows, the only classical thermodynamic parameters considered will be the temperature T, with the understanding that the pressure and chemical composition can be dealt with in basically the same way. Equation 3.2. can always be satisfied by the trivial solution:

$$G(T) = 0 \qquad\qquad\qquad 3.3$$

for one phase. This phase is conveniently taken as the high-symmetry phase and all quantities are measured with respect to this phase as **excess quantities**. The excess Gibbs free energy of the low-symmetry phase is then non-zero and, following the basic idea of Landau (e.g. Landau and Lifshitz 1980), we describe this excess free energy in an analytical form. The analytical form of the **Landau free energy or Landau potential** is then dictated by the symmetry of the high-symmetry phase and the symmetry breaking which leads to the low-symmetry form. Note that we are here extending the classical Landau theory rather dramatically: whereas Landau himself probably assumed in an ad hoc manner that the excess Gibbs free energy is well described in a Taylor series for *small values* of the order parameter, we assume here that the polynomial form of G is a good approximation over an extended temperature interval and that this approximation also holds for *large values* of the order parameter. We shall see that, indeed, values of $Q = 0.9$ can still follow 'Landau behaviour' although the original theory was not designed for these cases. It is a matter of taste whether or not this polynomial form of G is associated with the name of Landau and some authors seem to prefer the term 'Landau-like' or 'polynomial Gibbs free energy' in order to distinguish between Landau's perturbation approach and the full thermodynamic treatment. In the context of this book we shall use the term 'Landau potential' for all polynomial excess Gibbs free energies.

Before we discuss some simple examples it is useful to summarise some simple features of Landau theory which are relevant for their application within the context of ferroelastic and co-elastic materials. The excess enthalpy H of the crystal is mainly determined by the long-ranging elastic interactions which are a highly non-linear function of the order parameter. The entropy S, on the other hand, depends mainly on the energy dissipation due to phonon scattering and other mechanisms which are proportional to the pair correlation of the order parameter. The physical picture can be illustrated using the simplest forms of the excess quantities H and S :

$$S = -1/2\ AQ^2 \qquad\qquad\qquad 3.4$$

$$H = -1/2\ AT_c\ Q^2 + 1/4\ BQ^4 + 1/6\ CQ^6 + PV \qquad\qquad 3.5$$

As we are mainly concerned with the effect of temperature in this book, we shall ignore the effect of pressure P from the term PV (i.e. we set P=0) in what follows. If this effect is included, however, one normally finds that the excess volume V is proportional to Q^2 (see Chapter 8) so that PV μ PQ^2. This term is then similar to the entropy contribution μTQ^2. The external thermodynamic quantity pressure is then equivalent to temperature and all our future results equally well apply to pressure as to temperature if these two parameters were simply interchanged. The effect of chemical composition is similar to that of temperature and pressure as briefly discussed in Chapter 9.

If S and H are not explicitly temperature dependent, we can write the excess Gibbs free energy as:

$$G = H - TS = 1/2\,A\,(T\text{-}T_c)\,Q^2 + 1/4\,BQ^4 + 1/6\,CQ^6 \qquad 3.6$$

where all parameters A, T_c , B and C are independent of temperature. The Gibbs free energy in equation 3.6 if often referred to in common parlance as the 2-4-6 potential because of the contributions in the 2nd, 4th and 6th power of the order parameter. If the entropy is a non-harmonic function of the order parameter, as expected if configurational contributions are not fully described by Q^2, the parameters B and C also become explicit functions of temperature (see, for example, the instructive model calculations by Normand et al. 1990). Such non-harmonics of S may even be essential for the understanding of ferroelastic phase transitions involving orientational degrees of freedom (e.g. in the case of cyanides, Ohno (1987)). It is then often appropriate to formulate mean-field theories of the Bragg-Williams type which differ significantly from Landau theory when applied to the analysis of experimental observations at temperatures well below the transition point. As the vast majority of ferroelastic and co-elastic phase transitions appear to follow the predictions of Landau theory, however, we shall always assume in what follows that G has a simple polynomial form. As G has to fulfil all symmetry requirements of the material on a macroscopic level, we can easily apply group theoretical arguments to determine the actual analytical form of the Landau potential. The group theoretical aspects of Landau theory have been worked out in great detail over the last three decades with major contributions by Birman (1966), Cracknell (1974) Toledano and Toledano (1976,1977,1980,1982,1988), Stokes and Hatch (1988) and many others. Such symmetry restrictions do not apply, however, for non-symmetry breaking transitions or non-convergent ordering in which the high symmetry

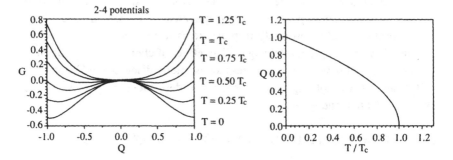

Fig.3.1 Landau potentials $G = \frac{1}{2}A(T\text{-}T_c)Q^2 + \frac{1}{4}BQ^4$ for various values of T. T is given in units of T_c. The minima occur at the equilibrium value of the order parameter Q for each temperature. The resulting temperature evolution of Q is plotted on the right hand side.

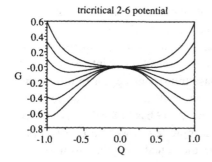

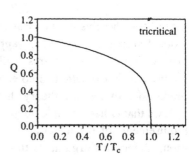

Fig.3.2 Landau potentials and order parameter behaviour for the tricritical potential $G = \frac{1}{2}A(T\text{-}T_c)Q^2 + \frac{1}{6}CQ^6$. Note the steeper decay of Q in the tricritical case compared with a second order transition in Fig.3.1.

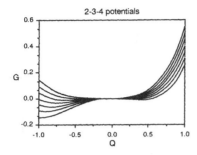
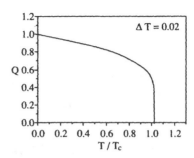

Fig.3.3 Landau potential and order parameter behaviour for a 2-3-4 potential
$G = \frac{1}{2} A(T\text{-}T_c) Q^2 + \frac{1}{3} BQ^3 + \frac{1}{4} CQ^4$. The parameters have been chosen to lead to a
transition at $T_{trans}=1.02\ T_c$. The phase transition is stepwise (i.e. first order) in
contrast to Fig.3.1 and Fig.3.2. Despite the stepwise character near T_{trans}, we find
a strong temperature dependence of Q at $T{<<}T_c$, similar to the tricritical case in
Fig.3.2. Note the assymetry in the Landau potentials due to the third order term.

form is only assymptotically approached. A common form of the Gibbs free
energy is then the 1-2-3 potential:

$$G = -HQ + \frac{1}{2} A (T\text{-}T_c) Q^2 + \frac{1}{3} BQ^3 \qquad\qquad 3.7$$

where H and B may also depend on temperature. In cubic, hexagonal and
trigonal systems, we often find a symmetry adapted Gibbs free energy in the
form of a 2-3-4 potential:

$$G = \frac{1}{2} A (T\text{-}T_c) Q^2 + \frac{1}{3} BQ^3 + \frac{1}{4} CQ^4 \qquad\qquad 3.8$$

which, in classical theory, always leads to a first order phase transition.

We finally mention the special case of the 2-4-6 potential in equation 3.6
where the fourth order term is zero (i.e. B=0) This transition is called
tricritical and represents the intermediate stage between continuous phase
transitions (B>0) and discontinuous transitions (B<0). The shapes of the
relevant Gibbs free energies are displayed in Fig.3.1-3.3. The thermodynamic
equilibria are determined by the minima of the potentials. The temperature
evolution of the minima determined thus the temperature dependence of the
order parameter, as shown in the Figures. The formal description of these

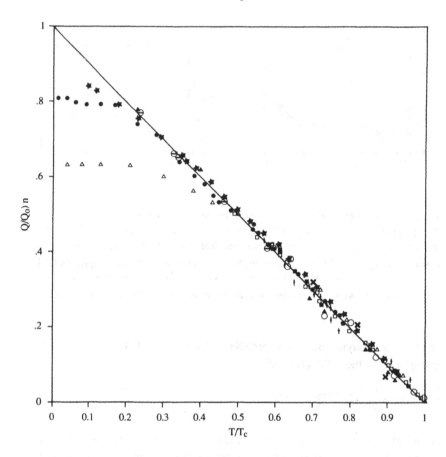

Fig.3.4 Temperature evolution of structural order parameters Q for some (nearly) second order and tricritical phase transitions. The temperature is normalised with respect to T_c the order parameter with respect to the extrapolated value at zero temperature, Q_0. [n = 2 for second order transitions: full circles: $LaAlO_3$, $T_c = 800K$, (Müller et al. 1968); open circles: $Na_{0.69}K_{0.31}AlSi_3O_8$, $T_c = 1251K$ (Salje 1968); circles with horizontal bar: $NaAlSi_3O_8$, $T_c = 251K$ (Grundy and Brown 1969); open squares: As_2O_5, $T_c = 578K$, (Salje et al. 1987, Redfern and Salje 1988); full squares: $Pb_3(AsO_4)_2$, $T_c = 560K$ (Bismayer and Salje 1981); open triangles: $Pb_3(P_{0.65}As_{0.35}O_4)$, $T_c = 455K$ (Bismayer et al. 1986); n = 4 for tricritical phase transitions: full triangles: $CaCO_3$, $T_c = 1250K$ (Dove and Powell 1989); dots with vertical bar: $CaAl_2Si_2O_8$, $T_c = 513K$ (Redfern and Salje 1988); stars: SiO_2, $T_c = 847K$ (Salje et al. 1989).

temperature dependencies will be derived later in this book. We give here only the most important solutions for a 2-4 potential:

$$Q^2 = \frac{A}{B}(T_c\text{-}T), \ T < T_c \qquad\qquad 3.9$$

and a tricritical 2-6 potential:

$$Q^4 = \frac{A}{C}(T_c\text{-}T), \ T < T_c \qquad\qquad 3.10$$

The general importance of these second order and tricritical phase transitions is demonstrated in Fig.3.4. Here the temperature evolution of the order parameters in different materials is plotted with a temperature axis in units of T/T_C where T_C is the transition temperature. The order parameters are normalised so that $Q=1$ at zero temperature. The exponent is $n=2$ for second order phase transitions and $n=4$ for tricritical phase transitions as given in equations 3.9 and 3.10. The Landau potentials would then predict that all measured values of Q for these phase transitions lie on one universal line. We can see from Fig3.4 that this prediction is backed up by the experimental observations for all temperatures which are not too close to absolute zero. At these very low temperatures, saturation of the order parameter occurs, which is discussed together with a microscopic derivation of the Landau potentials in the Appendix. Throughout this book we shall find further examples of the empirical observation that Landau potentials are an appropriate approximation for the excess Gibbs free energies of ferroelastic and co-elastic phase transitions.

4

THE SPONTANEOUS STRAIN

The central physical feature of ferroelastic and co-elastic crystals is their lattice distortion which is described by the 'spontaneous strain'. It is essential that the reader can deal with this quantity.

The order parameter and the Landau potential have been introduced in a fairly general manner without any restriction on ferroelastic or co-elastic phase transitions. The structural meaning of the order parameter Q can be related to the softening of an acoustic mode, an optical soft mode, an orientational ordering process, a cation exchange process or any other structural feature which is correlated with the phase transition. In order to make the phase transition ferroelastic or co-elastic, it is necessary that the transition mechanism changes the shape of the crystal in order to create a macroscopic spontaneous strain. We have seen that this macroscopic spontaneous strain is strongly influenced by the domain structure of the crystal. Although the domain effects are relevant to the macroscopic appearance of the mineral, they often do not contribute enough energy towards the total Gibbs free energy to influence the transition mechanism significantly. If microstructures are ignored as a first approximation, we can replace the macroscopic spontaneous strain by the **structural spontaneous strain** which is usually simply called the **spontaneous strain**. We shall come back to the question of microstructures in Chapter 7. This quantity is measured as the volume average of the structural deformation of the unit cell; its analytical form was first introduced by Aizu (1970) for ferroelastic systems. Since then it has been found useful to expand this definition to all structural phase transitions which lead to a variation of the shape of the crystallographic unit cell, in particular for co-elastic materials. If emphasis is given to the fact that the term spontaneous strain is used within the original context of Aizu's definition, the notation **Aizu strain** is used by some authors (see e.g. Wadhawan 1982).

The most common procedure for the experimental determination of spontaneous strain is to measure the temperature evolution of the lattice parameters using X-ray or neutron powder diffraction. A typical example of

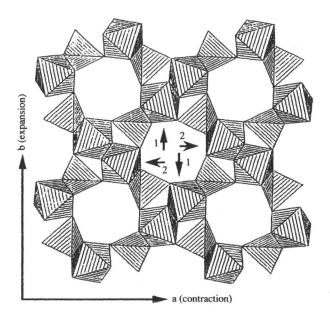

Fig.4.1 Crystal structure of As_2O_5 showing double-chains of octahedra interlinked by tetrahedra. The relevant structural element which is destroyed during the phase transition is the 4_1 axis along which the structure is projected in this drawing. The symmetry of this axis is reduced to 2_1 so that the two sets of octahedra indicated by the arrows 1 and 2 are no longer symmetry related. One type of chain tilts with respect to the other thereby contracting the lattice along the a-axis and expanding it along the b-axis. (Courtesy of W. Schmahl).

such experimental results is given by Redfern and Salje (1988) for the case of arsenic pentoxide, As_2O_5, (Fig.4.1). The crystal has tetragonal symmetry in the high-symmetry phase. This symmetry is reduced to orthorhombic in the ferroelastic phase with a characteristic change of lattice parameters from a=b at $T>T_c$ to b>a at $T<T_c$. The c-lattice parameter is not directly involved in the transition mechanism (i.e. a shear motion in the a-b plane) and no equivalent variation of the c-lattice parameter occurs. In order to evaluate the spontaneous strain, we have to extrapolate the lattice parameters of the high-symmetry phase into the temperature regime of the low-symmetry phase. This extrapolation represents that part of the thermal expansion which is not related to the structural phase transition and therefore does not contribute to the excess spontaneous strain. The numerical values of the spontaneous strain are now defined by the strain tensor which relates the low-symmetry unit cell to the high-symmetry unit cell when extrapolated to the same temperature.

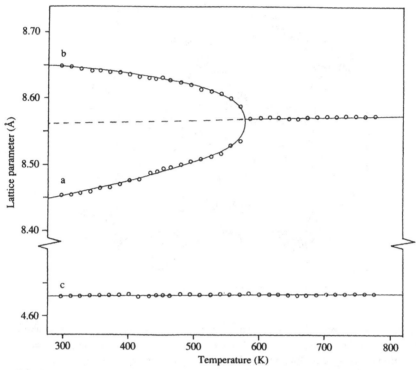

Fig.4.2 Temperature evolution of the lattice parameters in As_2O_5. The crystal has tetragonal symmetry at $T>T_c$ with a=b. At $T<T_c$ the symmetry reduces to orthorhombic b>a. The identical lattice parameters a and b in the high symmetry phase can be extrapolated into the temperature interval of the low symmetry phase (dotted line). The spontaneous strain is then related to the difference between the true lattice parameters and the dotted line which would have been obtained if the phase transition had not taken place.

In the case of arsenic pentoxide, the only non-zero components of the strain tensor are $e_{11} = - e_{22}$ which describe the expansion of the a-axis and the equivalent contraction of the b-axis. Note that since these two effects compensate for each other, the volume of the unit cell does not change (in linear approximation). The numerical values of e_{11} and e_{22} are given by:

$$e_{11} = - e_{22} = \frac{b_{orthorhombic} - b_{tetragonal}}{b_{tetragonal}}$$

4.1.

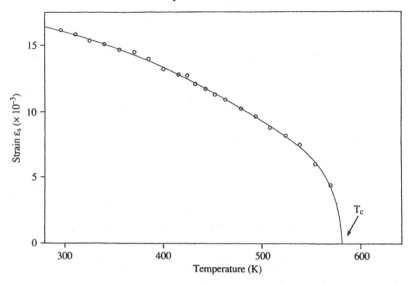

Fig.4.3 Temperature dependence of the spontaneous strain as derived from the lattice parameters shown in Fig.4.2. The line corresponds to the predicted behaviour of a second order phase transition using Landau theory.

where the tetragonal lattice parameter is always the extrapolated value at the temperature at which the orthorhombic lattice parameter is measured, see Fig.4.2. The resulting temperature evolution of the spontaneous strain is plotted in Fig.4.3. It clearly shows the transition temperature as the point at which the spontaneous strain disappears with the crystal being paraelastic at higher temperatures and ferroelastic at lower temperatures. By constructing the spontaneous strain we have reduced the total thermal expansion of the crystal to that part which is a true excess quantity (that is, it depends on the structural phase transition). The non-relevant background thermal expansion has been absorbed in the baseline in Fig.4.3.

The extrapolation of the lattice parameters of the high-symmetry phase over large temperature intervals in the low-symmetry phase obviously requires a good knowledge of the thermal expansion in the high-symmetry phase, which we shall assume here to be the high-temperature phase. As a rule of thumb it has been found that good experimental data have to be obtained over a temperature interval of at least 100K above the transition temperature. This can lead to considerable experimental difficulties in some mineral systems where the transition temperature is already very high and melting or chemical decomposition of the mineral occur at similar temperatures.Typical examples include calcite, with a transition temperature of 1250K and a CO_2 partial pressure of some 100bars at these temperatures, and in the case of Al,Si disordered Na-feldspar we find the transition temperature is near 1250K with melting occurring at around 1380K. Similar problems also occur at lower temperatures in $NaNO_3$ where the transition temperature at 551K is only 30K below the melting point (see, for extrapolation methods, Reeder et al. 1988). In all these cases it might be difficult to measure the thermal expansion at $T>T_c$ with sufficient accuracy to obtain a reliable extrapolation at $T<T_c$. It is common to use either of two approximations:

1 The thermal expansion of the high-symmetry phase is ignored and the values of the lattice parameters closest to the transition point are used as temperature-independent reference values for all temperatures (i.e. the dotted reference line in Fig.4.2 would be horizontal). This approximation might lead to systematic errors in the determination of the temperature evolution of the spontaneous strain, in particular if the numerical values of the spontaneous strain are small. A typical example is the $P\bar{1}$-$I\bar{1}$ phase transition in anorthite where the non-critical thermal expansion of the high temperature phase is comparable in magnitude with the spontaneous strain arising from the phase transition (Redfern et al. 1988). In most ferroelastic and co-elastic minerals with large spontaneous strains (say a few %) the systematic errors introduced by this procedure may be acceptable.

2 The extrapolated lattice parameters of the high-symmetry phase in the phase field of the low-symmetry phase are constructed using some structural model. The most common one is to define an 'average' structure (Wadhawan 1982). This structure is characterized by the arithmetic means of all possible strain tensors (i.e. for all domain orientations) in the low-symmetry phase. In arsenic pentoxide, for example, the two possible strain tensors are:

$$e_{ik} \text{ (S1)} = \begin{pmatrix} e_{11} & 0 & 0 \\ 0 & e_{22} & 0 \\ 0 & 0 & 0 \end{pmatrix} \qquad\qquad 4.2$$

and:

$$e_{ik} \text{ (S2)} = \begin{pmatrix} e_{22} & 0 & 0 \\ 0 & e_{11} & 0 \\ 0 & 0 & 0 \end{pmatrix} \qquad\qquad 4.3$$

These two tensors represent two structural states in which the a and b axes are interchanged. The average structure is defined by the mean value:

$$e_{ik} \ (av) = \begin{pmatrix} 1/2(e_{11}+e_{22}) & 0 & 0 \\ 0 & 1/2(e_{11}+e_{22}) & 0 \\ 0 & 0 & 0 \end{pmatrix} \qquad 4.4$$

The non-zero components of the latter tensor can now be expressed in lattice parameters, namely:

$$1/2(e_{11}+e_{22}) = 1/2 \ (\ (a-a_0)/a_0 \ + (b-b_0)/b_0) \qquad 4.5$$

and with $a_0 = b_0$ in the tetragonal phase, the condition that these tensor components vanish in the average structure (i.e. the hypothetical tetragonal structure at low temperatures) becomes:

$$a_0 = 1/2 \ (a+b) \qquad 4.6$$

Comparison with the experimental data in Fig.4.2 and 4.3 shows that this approximation is indeed very good. The second advantage of this approximation is that similar relationships can easily be worked out for all experimental situations without any information about the lattice parameters of the high-symmetry phase altogether.

The method has, on the other hand, a massive drawback. It is based on the assumption that the structural phase transition preserves the volume of the high-symmetry phase and that no other strain variations occur which are not fully described by the ferroelastic strain tensor (Aizu 1970). This assumption is clearly wrong for most co-elastic materials where volume changes are often encountered. More serious is the observation that sometimes even minerals which could reasonably be expected to be pure ferroelastics, such as leucite (Palmer et al 1989), do indeed show strong volume anomalies. In this particular case one finds that the construction of the spontaneous strain of leucite using the method of the average structure leads to a systematic error of almost 50% of the total spontaneous strain. These examples may illustrate that neither of these two approximations can be relied on without further detailed studies of the lattice parameters and, if possible, the thermal expansion of the crystallographic unit cell in the high-symmetry phase.

The spontaneous strain is a second rank tensor, and the magnitudes of its components are hence not uniquely definable because they depend on the setting of the two crystallographic unit cells. For any given setting, however, the tensor components can be determined directly from the measured lattice parameters. In the following equations we have denoted the lattice parameters of the low-symmetry phase as $a,b,c,\alpha,\beta,\gamma$ and those of the extrapolated high-symmetry phase as $a_0, b_0, c_0, \alpha_0, \beta_0, \gamma_0$. The most general formulation for a triclinic system can be put into the form:

$$e_{11} = \frac{a \sin \gamma}{a_0 \sin \gamma_0} - 1 \tag{4.7}$$

$$e_{22} = \frac{b}{b_0} - 1 \tag{4.8}$$

$$e_{33} = \frac{c \sin \alpha \sin \beta_0^*}{c_0 \sin \alpha_0 \sin \beta_0^*} - 1 \tag{4.9}$$

$$e_{23} = \frac{1}{2} \left(\frac{c \cos \alpha}{c_0 \sin \alpha_0 \sin \beta_0^*} + \frac{\cos \beta_0^*}{\sin \beta_0^* \sin \gamma_0} \left(\frac{a \cos \gamma}{a_0} - \frac{b \cos \gamma_0}{b_0} \right) \right)$$
$$- \frac{b \cos \alpha_0}{b_0 \sin \alpha_0 \sin \beta_0^*} \tag{4.10}$$

$$e_{13} = \frac{1}{2} \left(\frac{a \sin \gamma \cos \beta_0^*}{a_0 \sin \gamma_0 \sin \beta_0^*} - \frac{c \sin \alpha \cos \beta^*}{c_0 \sin \alpha_0 \sin \beta_0^*} \right) \tag{4.11}$$

$$e_{12} = \frac{1}{2} \left(\frac{a \cos \gamma}{a_0 \sin \gamma_0} - \frac{b \cos \gamma_0}{b_0 \sin \gamma_0} \right) \tag{4.12}$$

where the b axis has a common direction in both phases, and the z direction in the Cartesian system of the strain tensor is parallel to c* in the crystal (Redfern and Salje 1987). Stars indicate values of the reciprocal lattice parameters a*, b*, c*, α*, β*, γ* following the usual convention (see e.g. International Tables for X-ray Crystallography). Simplified expressions for higher symmetry phases follow from the eigen functions listed in Table 1 (see Appendix).

In the case of systems with higher symmetry the same formulae can be used, they simply lead to zero values for all symmetry forbidden components.

We have quoted the tensor components here in the standard tensor notation. Most authors do prefer the Voigt notation with:

$$e_1 = e_{11}$$
$$e_2 = e_{22}$$
$$e_3 = e_{33}$$
$$e_4 = 2e_{23}$$
$$e_5 = 2e_{13}$$
$$e_6 = 2e_{12}$$

4.13

Note that the tensor is symmetric and that the factors of 2 in these equations are often omitted in published data. These constant factors are irrelevant for any self-consistent theoretical treatment but they can lead to some confusion if absolute values of spontaneous strains are compared between different materials. For the definition of Aizu strains and the somewhat problematic use of the average structure as a reference state in the low-symmetry phase see Aizu (1970) and Wadhawan (1982).

Finally, it is convenient for a comparison of the degree of lattice distortion, described by the spontaneous strain in different systems, to define a scalar spontaneous strain $e_{spontaneous}$. The most common definition in the mineralogical and material science literature seems to be:

$$e_{spontaneous} = e_s = \sqrt{\Sigma\ e_i^2}$$

4.14

where e_i is the i-th component of the spontaneous strain in the Voigt notation (although this definition is only a convenience to convert strain into a scalar quantity, and is fairly arbitrary). Aizu (1970) used the tensor notation leading to the non-equivalent definition:

$$e_{Aizu} = \sqrt{\sum_{ik} e_{ik}^2}$$

4.15

Typical values of e_s for various minerals are listed in Table 2 (see Appendix).

5

COUPLING BETWEEN THE SPONTANEOUS STRAIN AND THE ORDER PARAMETER

Here we reconcile the thermodynamic quantity 'order parameter' and the geometrical quantity 'spontaneous strain'. Their temperature dependencies may follow each other but do not necessarily do so. Some symmetry arguments help to identify possible non-linearities. The coupling between the order parameter and the spontaneous strain produces an important physical effect: elastic softening of the crystal. Some simple algebra is needed to quantify the behaviour of the elastic constants. The reader is encouraged to work through this chapter because it illustrates some arguments which will be used repeatedly later on.

So far we have introduced the order parameter as the thermodynamic quantity which describes the phase transition and the macroscopic spontaneous strain as a convenient physical parameter which serves to characterise the change of shape of a crystal (or of the crystallographic unit cell for the structural or spontaneous strain). In some crystals, the order parameter and the spontaneous strain are simply proportional to each other. In most crystals this is not the case, however, and we must now explore rather carefully the possible correlations between the thermodynamic order parameter, Q, and the structural spontaneous strain, e_s. The general scope of this treatment is the **coupling theory** (e.g. Salje and Devarajan 1986, Achiam and Imry 1975, Gufan and Larin 1980, Imry 1975, Oleksy and Prysztawa 1983). It is based on the assumption that the crystal is in thermodynamic equilibrium with a surrounding heat bath and that its structural state is dictated by the condition that no structural variables can be changed without loss of energy. We shall furthermore assume in this Chapter that the crystal is homogeneous so that all fluctuation processes can be ignored.

The Gibbs free energy of a crystal with interacting Q and e_s can be formulated in three parts:

1 The Landau potential of the order parameter Q which we call L(Q);

2 The elastic energy from the relaxation of the unit cell which is described by the spontaneous strain, and the elastic constants of the paraelastic phase (i.e.

28

the 'bare' elastic constants), $1/2 \sum_{ik} C_{ik}e_ie_k$ where C_{ik} are elastic constants in Voigt notation, and;

3 The interaction energy between the order parameter Q and the spontaneous strain.

The Gibbs free energy can then be written as:

$$G(Q,e) = L(Q) + \frac{1}{2} \sum_{ik} C_{ik} \, e_i \, e_k + \sum_{mn} \zeta_{imn} \, e_i^m \, Q^n$$ 5.1

The last term represents the coupling energy in order n for the order parameter and in order m for the spontaneous strain (n,m >0) (ζ_{imn} are coupling constants between order parameters Q^n and strain elements e_i^m) . Just as for the order parameter Q and the spontaneous strain e_i, the coupling terms are subject to the constraints of symmetry and cannot be introduced arbitrarily. As a general rule, it holds that any combination of Q and e has to fulfil the same criteria to be a symmetry-allowed part of the Gibbs free energy as the higher order polynomial terms in a Landau potential. If, for example, the symmetry of Q and e are identical, we find that all combinations $Q^m e_i^n$ with n+m=p are allowed if Q^p is part of the Landau potential (this means in the language of group theory that Q and e_i transform according to the same irreducible representation).

Another general rule follows if e_i is a pure volume strain (i.e. transforms as the identity representation). In this case we find that all coupling terms m = 1, n > 1 are symmetry allowed, where Q^n is part of the Landau potential.

The simplest case is the so-called **bilinear coupling** term Qe_i. If bilinear coupling is allowed by symmetry, it follows automatically that all higher order couplings are also symmetry allowed. It was found by Devarajan and Salje (1984) that these higher order terms may contribute significantly to the strain energy in sulphates and it is likely that similar effects occur in other framework structures so that higher order coupling should not be automatically ignored even if bilinear coupling is compatible with symmetry.

No bilinear coupling can occur if the symmetry properties (i.e. the irreducible representations) of Q and e_i are different. The term which is always allowed by symmetry is $Q^2e_i^2$. This coupling term is called **biquadratic.** Further common coupling energies which have been observed in feldspars and langbeinites contain **linear-quadratic,** Q^2e_i and **linear-cubic,** Q^3e_i, terms. There is as yet little experimental evidence that any coupling of orders higher than m+n>4 provides a significant energy contribution and we restrict ourselves here to the discussion of the dominant bilinear and linear-quadratic coupling. The treatment of other coupling mechanisms is similar although analytically more strenuous (see Devarajan and Salje 1984, for the discussion of biquadratic coupling). Na-feldspar shows typical bilinear strain coupling behaviour, an example of linear-quadratic coupling is anorthite. As an example of higher order strain coupling we refer to Rocquet and Couzi (1985) who discuss the case of $Na_5Al_3F_{14}$.

5.1 Bilinear order parameter - strain coupling

The Gibbs free energy, in the case of bilinear coupling between the order parameter and the spontaneous strain, is in the most simple case of a Landau potential given by equation 3.6:

$$G(Q,e) = \frac{1}{2} A(T-T_c) Q^2 + \frac{1}{4} B Q^4 + \frac{1}{6} C Q^6 + \frac{1}{2} \sum_{ik} C_{ik} e_i e_k$$

$$+ \sum_i \zeta_i e_i Q \qquad\qquad 5.2$$

where the summation over i and k involves only those components of e_i which are allowed by symmetry (see Table 1, Appendix). Let us now consider a crystal which is in thermodynamic equilibrium with respect to the spontaneous strain. The correlation between Q and e_i then follows from:

$$\partial G (Q,e_i) / \partial e_i = 0 \qquad\qquad 5.3$$

and hence:

$$\frac{1}{2} \Sigma C_{ik} e_k + \zeta_i Q = 0 \qquad\qquad 5.4$$

which has the solution:

$$e_i = M_i \left(C_{ik}, \zeta_i \right) Q \qquad\qquad 5.5$$

where the coefficients M_i depend on the bare elastic constants C_{ik} and the coupling constants ζ_i. Replacing the strain variables in the Gibbs free energy by the order parameter leads to:

$$G(Q) = 1/2\, A(T-T_c^*)\, Q^2 + 1/4\, B\, Q^4 + 1/6\, C\, Q^6 \qquad\qquad 5.6$$

with the renormalized critical temperature:

$$T_c^* = T_c - 1/A \left(\Sigma \, C_{ik} \, M_i \, M_k + 2 \, \Sigma \, \zeta_i \, M_i \right) \qquad 5.7$$

As the strain energy is normally negative, this strain interaction leads to an increase of T_c and the stabilization of the low-symmetry phase.

We now explore the elastic behaviour of the crystal under the influence of external stress. If there were no structural instability described by the order parameter Q, then the elastic response would simply be determined from the elastic constants C_{ik}:

$$\sigma_i = \Sigma \, C_{ik} \, e_k \qquad 5.8$$

where e_k is now the strain induced by the stress σ_i. The role of the structural phase transition is to relax Q according to the external shape change. The condition that the strained crystal is in equilibrium with respect to the relaxation in Q is:

$$\partial G / \partial Q = 0 \qquad 5.9$$

at constant e_i, which leads, in linear approximation, to:

$$\frac{\partial^2 L}{\partial Q^2} Q + \sum_i \zeta_i \, e_i = 0 \qquad 5.10$$

The second derivative of the Landau potential with respect to the order parameter under the equilibrium condition $Q = Q_o$ indicates the curvature of the potential and is called the inverse order parameter **susceptibility** χ_Q^{-1}. This quantity is of utmost importance for scattering experiments and is widely discussed in literature (e.g. Lovesey 1987; Bruce and Cowley 1981). Equation 5.10 becomes then:

$$Q = - \Sigma \, \zeta_i \, \chi_Q \, e_i \qquad 5.11$$

The relaxation of the order parameter yields an additional reduction of lattice energy due to the Landau potential and the coupling energy. The effective elastic constants are smaller than without the lattice relaxation which leads to an **elastic softening** of the crystal. In the most simple approximation we can now write the Gibbs free energy as a function of the strain alone and find:

Salje

$$G = \frac{1}{2} \sum_{ik} C_{ik}^* \, e_i \, e_k + ...$$
5.12

with the renormalized elastic constants:

$$C_{ik}^* = C_{ik} - \zeta_i \, \zeta_k \, \chi_Q$$
5.13

Thus C_{ik}^* falls to 0 while χ_Q increases at the renormalised phase transition temperature T_c^*.

The relationship between the elastic constants and the order parameter as derived in equation 5.13 has been generalized by Slonczewski and Thomas (1970) for the interaction of several order parameters Q_m and arbitrary coupling. We can rewrite the coupling coefficients as:

$$\zeta_i = \frac{\partial^2 G}{\partial Q \partial e_i}$$
5.14

and the inverse susceptibility as:

$$\chi^{-1} = \frac{\partial^2 L}{\partial Q^2}$$
5.15

These relationships can now be generalised for several order parameters:

$$\zeta_{im} = \frac{\partial^2 G}{\partial Q_m \partial e_i}$$
5.16

and:

$$\chi_{mn}^{-1} = \frac{\partial^2 L}{\partial Q_m \partial Q_n}$$
5.17

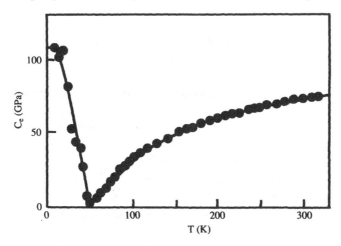

Fig.5.1 Influence of the structural phase transition on the elastic stiffness $C_e = \frac{1}{2}(C_{11}-C_{12})$ for Nb_3Sn (after Rehwald et al. 1972). The elastic softening at $T_c=50K$ can clearly be seen.

Its final form can be cast into the elegant formula:

$$C_{ik}^* = C_{ik} - \Sigma \ \frac{\partial^2 G}{\partial e_i \, \partial Q_m} \left(\frac{\partial^2 L}{\partial Q_m \partial Q_n} \right)^{-1} \frac{\partial^2 G}{\partial e_k \partial Q_n} \qquad\qquad 5.18$$

where the mixed derivatives of G indicate the coupling force constants and the second derivative of L is again the susceptibility of the Landau potential in the order parameter space. Typical examples for the temperature evolution of Nb3Sn, $BiVO_4$ and $LaNbO_4$ are shown in Fig.5.1 and 5.2. We shall discuss the detailed thermodynamic behaviour of systems with multidimensional order parameter spaces in Chapter 9.

We have seen so far that bilinear coupling between the order parameter and the spontaneous strain has the following consquences on the physical properties of the crystal:

1 The temperature at which the phase transition occurs for a stress-free crystal is renormalized with respect to the temperature T_C which appears in the bare Landau potential. The effect of temperature renormalization is very important in geological systems in which minerals occur within the matrix of a rock. The rather inhomogeneous elastic stress of this matrix then spreads the temperatures at which the mineral undergoes the phase transition over an interval which can be as large as several tens of degrees.

2 The temperature evolution of the order parameter can be followed in stress-free crystal via the temperature dependence of the spontaneous strain.

3 The elastic constants of the crystal depend explicitly on temperature if the order parameter and the strain can relax with respect to the external stress. The crystal will normally show an extensive softening of those elastic constants which are correlated with the spontaneous strain. In our present approximation we expect these elastic constants to disappear at the transition temperature T_C^* (see typical experimental results in Fig.5.1 and 5.2). A similar result would have been obtained if the physical origin of the order parameter was the elastic softening of the lattice itself, e.g. due to an anomaly of an acoustic phonon branch.

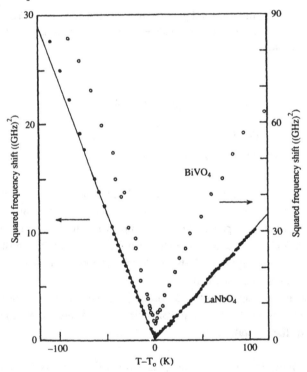

Fig.5.2 Elastic softening in $BiVO_4$ and $LaNbO_4$. The transition temperatures are T_C=528K for $BiVO_4$ and T_C=768K for $LaNbO_4$. The elastic constants are measured as squares of the phonon frequencies in a Brillouin scattering experiment (after Ishibashi et al. 1988).

In this case we can write the Gibbs free energy directly as:

$$G(e) = \frac{1}{2} \Sigma\, C_{ik}{}^* e_i\, e_k + \frac{1}{4} \Sigma\, C_{iklm}\, e_i\, e_k\, e_l\, e_m + \ldots \qquad\qquad 5.19$$

where the effective elastic constants $C_{ik}{}^*$ follow from equation 5.6 if we replace Q by the spontaneous strain via equation 5.5. It is $C_{ik}{}^*$ which now carries the full temperature dependence of the entropy term of the Landau potential.

It is rather unfortunate that the different mechanisms which lead to a temperature evolution of the elastic constants are often confused in published data. We have seen that $C_{ik}{}^*$ will behave as an inverse Landau susceptibility (i.e. $C_{ik}{}^* \propto |T-T_c{}^*|$) in the case of a direct acoustic instability. Experimentally, this behaviour is observed in Nb3Sn (Fig.5.1). The more common behaviour is related to a structural instability described by Q which induces elastic instability via a coupling mechanism. In this case we find that $C_{ik}{}^*$ is correlated with χ^{-1} and the coupling energies (equations 5.13 and 5.18).

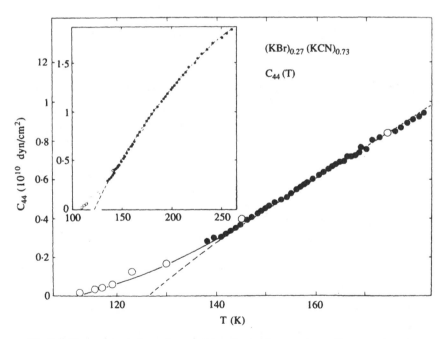

Fig.5.3 Temperature dependence of the elastic shear constant C_{44} (results of ultrasonic (filled circles) and inelastic neutron scattering experiments (open circles)). The dotted line is a best fit of the experimental data using equation 5.20 (after Knorr et al. 1986).

The resulting temperature anomalies are restricted to smaller temperature intervals than in the case of direct proportionality between $C_{ik}*$ and $|T-T_c*|$. Typical examples are $KBr_{0.27} KCN_{0.73}$ (Fig.5.3) and KH_2PO_4(Fig.5.4). Here it is almost irrelevant which physical mechanism underlies the structural instabilies (Elliott et al. 1971). The experimental results of Sandercock et al. (1972) on the Jahn-Teller phase transition in $TbVO_4$ in Fig. 5.5 can be well described by equation 5.13 .

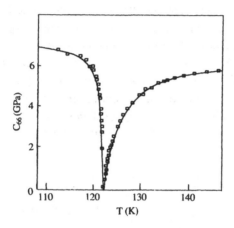

Fig.5.4 Elastic softening of the elastic shear constant C_{66} in KH_2PO_4 (after Garland and Novotny 1969).

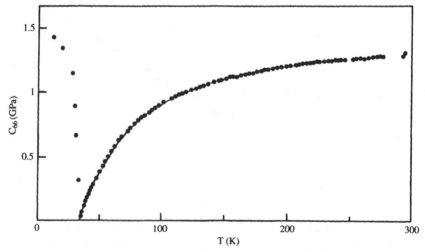

Fig.5.5 Ultrasonic (T>Tc) and Brillouin scattering (T<Tc)results for the temperature dependence of C_{66} in $TbVO_4$ (after Sandercock et al. 1972).

Following the same line of argument, Feile et al. (1982) propose the general expression:

$$C^* = C\,\frac{T\text{-}T_0}{T\text{-}T_a}$$

5.20

which is a simple parametrisation of e.g. 5.18. χ^{-1} vanishes at T_a and C^* extrapolates to zero at T_0. This form was used by Knorr et al. (1986) to fit their experimental data in Fig.5.3.

We now return to the discussion of the coupling terms in equation 5.18. We have seen that the mixed derivatives $\partial^2 G\,/\partial e_i\partial Q_m$ are independent of the order parameter and the spontaneous strain but this does not mean that these quantities must be independent of temperature. As they have the same symmetry as the entropy in simple Landau theory, we would even expect them to show at least a mild temperature dependence.

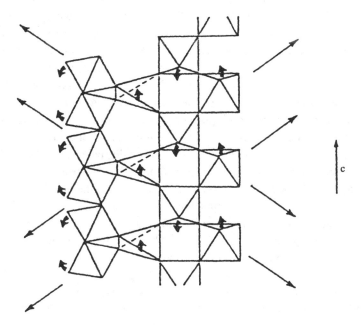

Fig.5.6 Atomic displacement pattern during the tetragonal → orthorhombic phase transition in As$_2$O$_5$. Fine arrows indicate the projections of the polyhedra tilt axis which lie in the {110} planes (after Schmahl and Redfern 1988).

To illustrate the basic results of this Chapter further, we return to As$_2$O$_5$ as a ferroelastic material with bilinear order parameter-strain coupling. The crystal structure is shown in Fig.4.1. It consists of a framework of octahedral chains interlinked with tetrahedral complexes (Jansen 1977). The room temperature orthorhombic form, space group $P2_12_12_1$ transforms to the tetragonal phase space group $P4_12_12$, above 578K. The active irreducible representation is B_1. Bismayer et al. (1986) described the relevant part of the Gibbs free energy for the phase transition in terms of a Landau potential of the order parameter, the elastic energy and the bilinear coupling between the order parameter and the spontaneous strain (e_1-e_2):

$$G = \frac{1}{2} A(T\text{-}T_c) Q^2 + \frac{1}{4} BQ^4 + \zeta\, Q\,(e_2\text{-}e_1) + \frac{1}{16} C^*(e_2\text{-}e_1)^2 \qquad 5.21$$

where C^* is the approximate combination of elastic constants.

Salje et al. (1987) and Redfern et al. (1988) have shown that this free energy expression does indeed correctly describe the temperature evolution of the spontaneous strain and the morphic birefringence over a large temperature interval between room temperature and T_c. The structural changes were analysed by Schmahl and Redfern (1988) using X-ray scattering methods. These authors found that the relevant changes of the crystal structure during the phase transition are due to the loss of the screw periodicity of the structure along the c-axis. Atoms within one unit cell are displaced out of phase parallel to the c-axis in the low-symmetry phase with a displacement pattern of optic rather than acoustic character (see also Fig.5.6). Additional octahedra tilts around axes within the {110} planes lead to atomic displacements with a component parallel to the c-axis. As these movements are antiparallel to each other, they display also an optical character of the displacement pattern. The equivalent softening of Raman active phonon branches has been observed by Salje et al. (1987). Arsenic pentoxide is thus an example of a ferroelastic material which shows a large spontaneous strain ($e_s = [a\text{-}b]/a_0 = 0.023$ at room temperature) and a structural order parameter which primarily describes the distortion of polyhedral complexes. The driving force of the phase transition is consequently not simply due to the gain of elastic energy in the low-symmetry phase but is closely related to the gradual softening of optical phonons together with their strong interaction with the acoustic modes. The predicted temperature evolution of C_{ik}^* has not yet been investigated experimentally, because of a lack of suitably sized crystals.

5.2 Linear-quadratic coupling between strain and order parameters

At all structural phase transitions, the order parameter can couple with the elastic strain via a coupling energy that is linear in the strain and quadratic in the order parameter. This interaction energy may be sufficiently small to be ignored, as in many magnetic and polytypic systems, and in liquid crystals. In all co-elastic materials this interaction is, by definition, relevant for the understanding of the phase transition mechanism and hence forms an important part of the Gibbs free energy. In order to illustrate the physical picture, we start again from the simplest Landau potential in equation 3.6 and write the Gibbs free energy including the elastic energy and the coupling energy as:

$$G(Q,e) = \frac{1}{2} A(T\text{-}T_c) Q^2 + \frac{1}{4} B Q^4 + ... + \frac{1}{2} \sum_{ik} C_{ik} e_i e_k + \sum_i \zeta_i Q^2 e_i \qquad 5.22$$

The condition for the material to be stress-free is again:

$$\frac{\partial G}{\partial e_i} = 0 \qquad 5.23$$

leading to the correlation between the spontaneous strain and the order parameter:

$$-Q^2 \zeta_i = \frac{1}{2} \Sigma \; C_{ik} e_k \qquad 5.24$$

with the solution:

$$e_i = M_i \left(C_{ik}, \zeta_i \right) Q^2 \qquad 5.25$$

This relationship between e_i and Q is similar to the bilinear case in equation 5.1.4 with the significant difference that the spontaneous strain now reflects the square of the order parameter rather than the order parameter itself. In a classical second order phase transition with $Q \propto |T\text{-}T_c|^{1/2}$ we would hence expect that the spontaneous strain should evolve linearly with temperature, as observed in the displacive phase transition in anorthite (Fig.5.7). We find the same relationship also for crystals in which bilinear coupling is 'symmetry allowed' for some components of the spontaneous strain tensor. All other

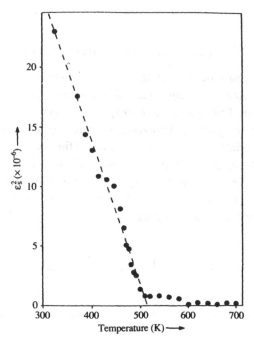

Fig.5.7 Variation with temperature of the square of the spontaneous strain, which is proportional to Q^{o4} (anorthite, Val Pasmeda locality). The linear dependence observed indicates $Q^o \propto (T^*_c-T)^{1/4}$ and reflects the tricritical nature of the phase transition.

components of this tensor can then couple linear-quadratically with the order parameter, i.e. all components of the strain tensor with the same symmetry as the order parameter couple bilinearly, whereas all other components couple linear quadratically. The two sets of components show, thus, different temperature dependencies. To illustrate this effect let us briefly return to our example of As_2O_5. Bismayer et al. (1986) have investigated the non-linear coupling between the order parameter and strain components with symmetry different from that of the order parameter. It was found that this coupling leads to the formation of the strain components e_3 and e_6 (Fig.5.8), which reflect the square of the order parameter whereas e_1 and e_2 are directly proportional to the order parameter (Fig.4.2 and 4.3).The Gibbs free energy follows from substituting 5.25 into 5.22:

$$G(Q) = \frac{1}{2} A(T-T_c) Q^2 + \frac{1}{4} B^* Q^4 + ...$$

$$5.26$$

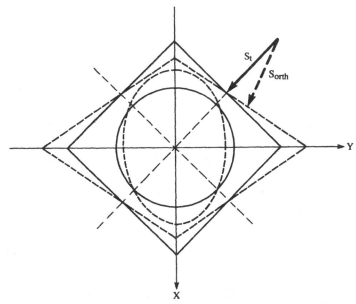

Fig.5.8a) Distortion of an As_2O_5 crystal due to the structural phase transition. The spontaneous strain is $e_2-e_1 \propto Q$ which leads to an elongation along the Y axis and a contraction along the X axis. This movement is proportional to the order parameter. As a consequence, the lattice planes {110} are rotated and their normals are changed from S_t to S_{orth}. This rotation changes the distance between the lattice planes which is proportional to the cosine of the rotation angle (i.e. e_6). As the rotation angle is itself proportional to the spontaneous strain (e_2-e_1) and $\cos \psi = 1 - 1/2\ \psi^2$, $\psi \propto Q$ we find that $e_6 \propto Q^2$. Inserted is the projection of the deformation ellipsoid on the (001) plane. which results from the deformation at $T<T_c$ of a sphere at $T>T_c$.

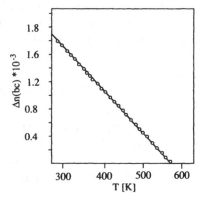

Fig.5.8b) Experimentally observed temperature evolution of the optical birefringence which is proportional to e_6. Note the linear temperature dependence of e_6 ($\propto Q^2$) which is in contrast to the root-function of e_1-e_2 ($\propto Q$) in Fig.4.3.

with:

$$B^* = B + 2\Sigma \; C_{ik} \; M_i \; M_k + 4\Sigma \; \zeta_i \; M_i \qquad\qquad 5.27$$

One can easily show for the case of high symmetry systems that the last two terms have opposite signs and that their sum is negative. The strain interaction again leads to a renormalised Landau potential with a fourth order coefficient that is smaller than without the strain interaction. If this renormalization is strong enough it can lead to a phase transition being first order even if it would be second order under constant strain conditions.

The elastic constants of the crystal are normally measured in thermodynamic equilibrium with:

$$C_{ik}^* = C_{ik} - \frac{\partial^2 G}{\partial Q^2} \frac{\partial Q}{\partial e_i} \frac{\partial Q}{\partial e_k} \qquad\qquad 5.28$$

$$= C_{ik} - B/(12 \; M_i \; M_k) - A(T-T_c)/\left(4 \; M_i \; M_k \; Q^2\right) + ... \qquad 5.29$$

The first term C_{ik} is the bare constant which would be measured without coupling between the strain and the order parameter. Experimentally, this could happen if a local stress field is applied at high frequencies so that the elastic deformation can follow this stress field but not the order parameter in cases where the relaxation process of the order parameter is sufficiently slow. The renormalization due to the second term in equation 5.29 leads, at lower experimental frequencies, to a jump of the elastic constants, even when the phase transition itself is continuous. The last term signifies a subsequent explicit temperature dependence of the elastic constants in the low-symmetry phase for all phase transitions except simple second order ones where this term indicates an additional constant contribution to the jump of C_{ik}^*. Rehwald (1973) has shown that the elastic constants of many co-elastic and ferroelastic crystals with strong linear-quadratic coupling between the spontaneous strain and the order parameter indeed follow the predicted behaviour. A quantitative analysis of experimental observations in oxides and framework structures show an excellent agreement with the predictions of Landau theory (e.g. the review of Lüti and Rehwald 1981), a typical example of the elastic anomalies in the lead phosphate is shown in Fig 5.9 (after Torres et al. 1982a,b).

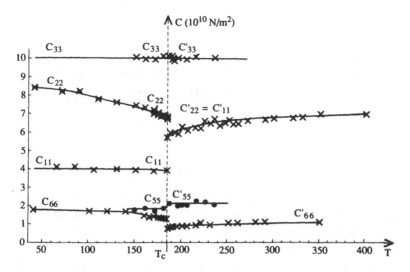

Fig.5.9 Temperature evolution of the elastic constants in $Pb_3(PO_4)_2$ from Brillouin
scattering experiments. The strain couples linear-quadratically with the order
parameter and the elastic constants show a stepwise behaviour at the transition point
with some weak additional elastic softening over a large temperature interval.
Note the difference between this behaviour and the strong softening in the case of
bilinear coupling in Fig.5.1 to 5.5 (after Torres et al. 1982a).

6

MACROSCOPIC CLASSIFICATION OF FERROIC AND CO-ELASTIC CRYSTALS

Some common definitions: necessary, useful but very dry.

Phase transitions accompanied by a change of the point-group symmetry are called **ferroic**. Ferroic phase transitions are often subdivided into **ferrodistortive**, i.e. where the translational symmetry is preserved during the phase transition, and **antiferrodistortive** where such an invariance does not exist. In the language of group theory it means that the active irreducible representation has Γ- point symmetry (k=0) in the case of a ferrodistortive phase transition, whereas it transforms with a finite k-vector for all other transitions. This k-vector is called **critical**: the critical k-vector is at special points of the boundary of the first Brillouin Zone for antiferrodistortive phases and inside the Brillouin Zone, but not at its origin, for modulated phases. Various aspects of these symmetry relationships were discussed within the framework of Landau theory by Toledano and Toledano (1980), Stokes and Hatch (1988), Janovec (1976) and many other authors. These symmetry considerations are useful for classification purposes, but they do not necessarily always help to elucidate the underlying physical properties which actually drive the transition mechanisms. The reason for this is that, usually, several physical processes combine to lower the Gibbs free energy so that the low-symmetry phase becomes thermodynamically stable. Typical examples are combinations between electric and elastic degrees of freedom (e.g. Suzuki and Ishibashi 1987), molecular ordering and co-elasticity (Lynden-Bell et al. 1989) and cation ordering and ferroelasticity (e.g. Salje 1985; and many others). It is obvious that any meaningful analysis of the transition behaviour must incorporate these coupled processes which are described in more detail in Chapter 10.

In this Chapter we shall discuss under what circumstances coupling phenomena may lead to macroscopic ferroic behaviour of the low-symmetry phase and will classify ferroic materials according to their physical behaviour.

We start in the spirit of the treatment of ferroic state shifts by Aizu (1973) and write the macroscopic Gibbs free energy of a co-elastic crystal as:

$$G = L(Q) + \Sigma \, e_i \, \sigma_i + \Sigma \, \zeta_{imn} \, Q^m \, e_i^n + \ldots \qquad \qquad 6.1$$

where L(Q) again indicates the Landau potential in Q. The parameter e_i now expresses the total strain of the low-symmetry phase including the possibility of creating strain by external forces. The total strain is identical with the spontaneous strain if such external forces do not exist and equation 6.1 becomes identical to equation 5.1. The external forces are described by fields, in particular the stress field σ, the electrical field E and the magnetic field H'. The total strain is then:

$$e_i = (e_s)_i + \sum s_{ik}\sigma_k + \sum d_{ik}E_k + \sum q_{ik}H' \qquad \qquad 6.2$$

The coefficients are components of the elastic compliance tensor, s_{ik}, the piezoelectric tensor, d_{ik}, and the piezomagnetic tensor q_{ik}. We are using the Voigt notation throughout. Each of the parameters $(e_s)_i$, s_{ik}, d_{ik} and q_{ik} can couple with the order parameter in the same way as discussed before in the case of spontaneous strain. Each of these coupled quantities will then show an excess behaviour due to the structural phase transition. The coupling will also lead to the formation of domain structures, either potentially or as observed under suitable experimental conditions, where each domain has a different value of the total strain parameter. Those parameters which couple with the order parameter will participate in the domain structure, those which are independent of the order parameter will stay homogeneous. A suitable classification can now be obtained if we consider the difference in Gibbs free energies between domains with the structural states S_1 and S_2. Let G_1 be the Gibbs free energy of state S_1 and G_2 that of the state S_2. The difference between G_1 and G_2 is:

$$\Delta G = \sum \Delta(e_s)_i \sigma_i + \sum \Delta s_{ik}\sigma_i\sigma_k$$

$$+ \sum \Delta d_{ik}E_i\sigma_k + \sum \Delta q_{ik}H'_i\sigma_k \qquad \qquad 6.3$$

where the Δ sign indicates which parameter changes during the state shift. The following terms are now commonly used to describe the four possible state changes:

$\Delta(e_s)_i$	ferroelastic (primary ferroic)
Δs_{ik}	ferrobielastic (secondary ferroic)
Δd_{ik}	ferroelastoelectric (secondary ferroic)
Δq_{ik}	ferromagnetoelastic (secondary ferroic)

Strictly these terms should be used only if the switching process under the relevant field has been observed experimentally, otherwise the qualifying term **potential** should be added as in the case of ferroelastics. If no two of the possible states S_1, S_2, in the low-symmetry form have all their corresponding non-zero components of $(e_s)_i$, s_{ik}, d_{ik} or q_{ik} identical, then the crystal is called **full** ferroelastic, ferrobielastic, ferroelastoelectric or ferromagnetoelastic, respectively (Wadhawan 1982). In all other cases the adjective **partial** is used. For the definition of other, non co-elastic, higher order ferroics, see the review of Wadhawan (1982). Equivalent definitions exist for electric ferroics in which the strain is replaced by the dipole moment. The primary ferroic is then ferroelectric. Details of ferroelectric and related compounds have been compiled by Lines and Glass (1977).

Besides the definition of the excess parameters during the co-elastic phase transition it is also rather useful to classify the various degrees of coupling between the order parameter and the excess quantity. Following the usual convention we define all those effects due to a coupling which is linear in order parameter as **proper** and all those which are non-linear as **improper**. A similar definition is, of course, widely used for proper and improper ferroelectrics and ferromagnets although we shall see that in the case of ferroelastic minerals the more common behaviour is related to improper ferroelasticity (e.g. the palmierite structure) whereas there are only a few known cases where ferroelasticity occurs with bilinear coupling between the order parameter and the spontaneous strain (e.g. As_2O_5). Finally we mention the common definition which distinguishes between **pure** and **impure** ferroelastics according to whether the ferroelastic phase transition occurs as the only phase transition of the material (pure) or whether other phase transitions also occur. The distinction between pure and impure ferroelastics appears to be particularly useful in the area of Materials Science because it indicates in the case of impure ferroelastics the possible influence of dielectric or magnetic instabilities which might make the device application of ferroelastic materials difficult. From a physicist's viewpoint it seems, however, more desirable to distinguish crystals in which the ferroelastic effect is driven entirely by an elastic instability (e.g. via an acoustic soft mode) from those crystals in which ferroelasticity occurs as a secondary effect. The more general definition which encompasses both ideas is that a ferroelastic crystal is called pure if there is only one order parameter involved in the phase transition. The ferroelastic is called impure if several order parameters representing different transition mechanisms are associated with the phase transition. Details of the coupling between different order parameters is discussed in detail in Chapter 10.

Some typical examples of ferroelastic crystals are given in Table 3 below.

Table 3 Classification of ferroelastic and co-elastic phase transitions and some examples used throughout this book

	proper ($Q \propto e$)	improper ($Q \not\propto e$)
pure {Q}	As_2O_5 (if rotations in Fig.5.6 are ignored) $NaAlSi_3O_8$ (Al,Si disordered Na-feldspar)	$Pb_3(PO_4)_2$
impure $\{Q_1, Q_2 ...\}$		$CaAl_2Si_2O_8$ anorthite $KAlSi_2O_6$ leucite

7

FERROELASTIC AND CO-ELASTIC TWIN STRUCTURES

The static twin structures are explained here: which twins occur and how they interact. The major microstructures are: rounded junctions where twins intersect, S-shaped walls, and needle twins. Phenomenological description and theory are separated in these chapters. If the reader is only interested in the descriptive part, he or she can ignore 7.1, 7.2b, 7.3b and 7.4b. To help the reader, a summary of the main ideas discussed, is provided at the end of each theoretical section.

Twins in minerals are formed by several mechanisms, which according to Burger (1945), can be classified into growth, transformation and gliding. In the absence of external fields, as described in the last chapter, transformation twins are entirely due to the generation of the spontaneous strain during a co-elastic phase transition. Their geometrical and physical properties then follow directly from the temperature evolution of the spontaneous strain and we shall develop a concept which allows us to identify the essential features of twin boundaries on a length scale which is still rather unaffected by direct structural properties (i.e. we shall use the continuum approximation on a mesoscopic length scale). In this Chapter we shall treat such twin structures as a static phenomenon, expected to occur at temperatures not too close to the transition point. Some aspects of the dynamics of the domain structure are explored in Chapter 8.

It is rather useful at the beginning of the discussion of twin structures and their dynamics to recall the surprisingly large strain and stress fields which exist in a co-elastic material. A typical order of magnitude for a spontaneous strain in a framework mineral is 2%, or, in the language of metallurgy, the aspect ratio of a mineral is changed by some 2% during a structural phase transition. The correlated changes in the enthalpy of the crystal related to this formation of spontaneous strain often reach some 6 kJ/mole, an energy which would lead to changes in thermochemical or petrological phase diagrams of some hundreds of degrees in temperature and Giga-pascals in pressure. The energy content of the co-elastic material due to the structural phase transitions is thus rather large and is certainly relevant to the thermodynamic behaviour of the material. Furthermore, the internal stresses which can build up due to specific constellations of twin structures are of the same order of magnitude as the stress which is necessary to remove the spontaneous strain. With $e_S = 0.02$ and an effective elastic constant of 30 GPa we find that this stress can be as large as 0.6 GPa. It is not surprising that we shall find that lattice planes in a crystal are easily bent and large proportions of the bulk of a crystal heavily warped under the influence of twin structures.

A static domain wall between two adjacent twins can be envisaged as an internal surface of the crystal. The orientation of an individual twin wall is then determined by the condition that the crystal in the low-symmetry phase tends to maintain the total symmetry of the high-symmetry phase as a statistical average. Symmetry elements such as diads and mirror planes which are destroyed during the structural phase transition can give rise to the formation of twin boundaries. This criterion has been used by Dvorak (1978) to anticipate the symmetry of the high-symmetry phase from the observed twin-structure in the low-symmetry phase; even if such a phase is hypothetical because the crystal decomposes or melts under heating before the actual high temperature phase is reached. The total number, N, of domains in the low-symmetry form as created by the structural phase transition from the high-symmetry form can easily be anticipated from the following criterion:

$$N = \frac{\text{(number of symmetry elements in the high-symmetry phase)}}{\text{(number of symmetry elements in the low-symmetry phase)}} \qquad 7.1$$

The group theoretical argument behind this criterion is as follows. Let G_0 be the point group of the high-symmetry phase and G the point group of the low symmetry phase. Following the usual convention we call the groups G and G_0 and trust that the reader will not confuse them with the Gibbs free energy. In Landau theory we assume that G is a subgroup of G_0 and we can write G_0 as a finite series of co-sets:

$$G_0 = G + R_1* G + R_2*G + ... + R_n*G \qquad 7.2$$

where R_1 .... R_n are elements of G_0 which do not occur in G. Each co-set represents one type of domain with the orientation between the crystallographic axes of the domains determined by the elements of the co-set. As the total number of elements G_0 is finite and the series is complete, we find that the number of co-sets is the number of elements in G_0 divided by the number of elements in G.

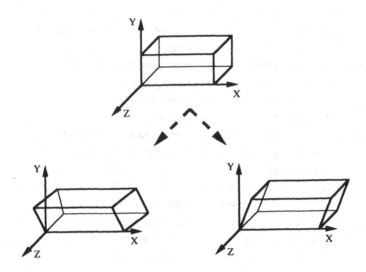

Fig.7.1 Shear deformation in NdP_5O_{14} (mmm → 2/m). The shear strain is chosen as $e_6 = 2e_{xy}$. Two domain orientations are possible with the strain $+e_6$ and $-e_6$.

Let us illustrate this argument with the example of the ferroelastic phase transition in NdP_5O_{14} (T_C=419K) in Fig.7.1. The high-symmetry form is paraelastic with the point-group symmetry mmm (D_{2h}) and the low-symmetry phase is ferroelastic with the monoclinic point-group symmetry 2/m (C_{2h}). Without knowing the details of the crystal structure, we can immediately infer the domain structure from the symmetry relationship:

$$\{mmm\} = \{ 2/m \} + m(x,y) * \{ 2/m\} \qquad\qquad 7.3$$

where the curved brackets indicate the point groups and m(x,y) is one of the two mirror planes which exists in {mmm} but not in {2/m}. The total number of domains is two because there are two co-sets. This follows also from the fact that there are eight elements in mmm and four elements in 2/m so that N = 8/4. The boundary between the two domains is a twin boundary because it is related to the mirror plane m(x,y), the orientation of the twin walls is (101) and (001) (see Fig. 7.1).

The number of possible domains is listed for ferrodistortive ferroelastics and co-elastic materials in Table 1. The energy of such a domain wall can be written as follows:

$$G(\text{wall}) = \int L\,(Q)\,dV + \Sigma \int (e_s)_i\,\sigma_i\,dV + \int \sigma_w\,dS$$

7.4

where L is the energy density due to the Landau potential on a mesoscopic scale for areas in the crystal which are close to the domain wall. The second term indicates the elastic interaction between the spontaneous strain and a stress field and the last term represents the surface energy of the domain wall.

The energy contribution due to the order parameter on a mesoscopic scale is now illustrated by using an Al,Si disordered Na-feldspar as an example. A structural phase transition takes place in this mineral between a monoclinic phase C2/m and a triclinic phase $C\bar{1}$ (Fig.7.2). The translational symmetry is not changed during the phase transition, the critical k-vector is the Γ-point (i.e. k=0) and the transition is thus ferrodistortive. The elastic part of the Gibbs free energy accounts for about 50% of the total excess energy (Salje et al. 1985a) which allows us to call this phase transition co-elastic. The order parameter, Q, is related to the tilt of the Al,Si - O crankshafts (Fig.7.2) which, in thermodynamic equilibrium, couples with the second order parameter of the Al,Si ordering, Q_{od} (Salje 1985, Salje et al. 1985b). Twin walls are formed during the phase transition, with the twin plane perpendicular to the pseudo-diad, leading to what we call the Albite twin law. The second twin law is the Pericline law, where the twin wall is parallel to the pseudo-diad (e.g. Goldsmith and Laves 1954). Part of the twin wall energy is now related to the variation of the local order parameter due to the formation of the twin boundary. In the case of fully Al,Si disordered Na-feldspar, this energy is zero provided no local order is induced by the domain wall. In more ordered Na-feldspar, this is no longer true because a Pericline wall will always induce local Al - O - Al bonds and thereby increase the energy of the domain wall. We shall absorb this energy contribution in the last term of equation 7.4, whereas the elastic deformation of the lattice planes close to the wall contributes to the first two terms.

We will now discuss the strain part of the wall energy. It is clear that even when no external stress is applied to the crystal, the strain energy will be finite because two adjacent domains are locally stressed in order to join along the wall. The criterion for the wall orientation and the shape of walls when interacting, results mainly from the minimization of this strain energy. Simple strain compatibility criteria based on this approach were worked out by Fousek and Janovec (1969) and Janovec (1976): for their application see e.g. Balagurov et al. (1986). An elegant formulation was introduced by Sapriel (1975) and was subsequently used in the mineralogical context for the

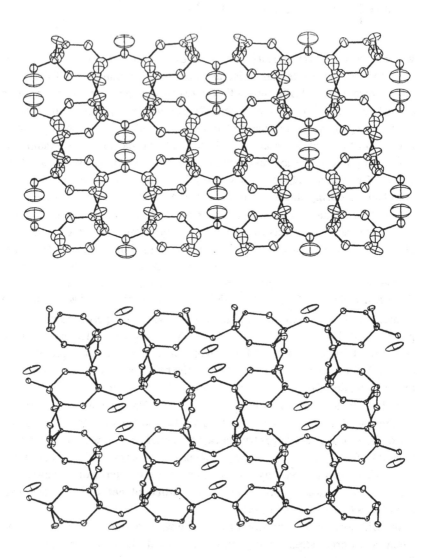

Fig.7.2 Crystal structures of Na-feldspar above and below the transition point. The upper graph shows the monoclinic form with vertical mirror planes which go through the large Na-positions (C2/m). These mirror planes do not exist in the lower graph which shows the triclinic structure (C$\bar{1}$). The structural 'crankshafts' of Al-Si-O tetrahedra extend from the top to the bottom. In the monoclinic form, they are vertical and related by symmetry. In the triclinic form, the crankshafts have rotated around their axes parallel to the mirror planes.

analysis of domain pattern of feldspars (Salje et al. 1985a) and leucite (Palmer et al. 1988,1989). It is based on the condition that the spontaneous strains of two adjacent domains $e_s(S_1)$ and $e_s(S_2)$ must be identical in the domain boundary. Consequently the differences between the respective strain components must vanish and the surface is determined by the condition:

$$(e_{ik}(S_1) - e_{ik}(S_2)) \, x_i \, x_k = 0 \qquad\qquad 7.5$$

in tensor notation where x_i (i=1,2,3) denotes the Cartesian coordinates of a point in the domain wall, measured with respect to an origin on the plane. This condition splits into a product of two linear equations if the difference between the two strain tensors is traceless. This condition is always fulfilled if the tensor of the spontaneous strain is itself traceless which means, in geometrical terms, that the volume of the crystal is not changed during the phase transition. This condition is met by all proper ferroelastics as shown in Table 1 if we ignore any linear-quadratic coupling between the volume and the order parameter. Most other ferroelastics show small volume anomalies and it appears that the excess volume contraction of leucite of 4% is an extreme case (see Palmer Chapter 21). Large volume anomalies are, however, common in non-ferroelastic co-elastic minerals such as calcite (other anomalies include $Cd_2 (NH_4)_2(SO_4)_3$ and $Cd_2Tl_2 (SO_4)_3$, Sapriel 1975 and H_3BO_3, Wadhawan 1978).

Solutions of equation 7.5 give two permissible domain walls which are perpendicular to each other if there is no volume anomaly and if no interaction occurs between domain walls of different orientations. These two types of walls are called W-walls if they are crystallographic planes of fixed indices with respect to the high-symmetry phase. They are called W' walls in all other cases (Sapriel 1975).

Let us illustrate this distinction between W and W' walls in feldspars. The spontaneous strain tensors of two adjacent domains in triclinic feldspar are:

$$e_{ik} (S_1) = \begin{pmatrix} 0 & e_{12} & 0 \\ e_{12} & 0 & e_{23} \\ 0 & e_{23} & 0 \end{pmatrix} = (0\ 0\ 0\ e_4\ 0\ e_6) \qquad\qquad 7.6$$

and

$$e_{ik} (S_2) = \begin{pmatrix} 0 & -e_{12} & 0 \\ -e_{12} & 0 & -e_{23} \\ 0 & -e_{23} & 0 \end{pmatrix} = (0\ 0\ 0\ -e_4\ 0\ -e_6) \qquad\qquad 7.7$$

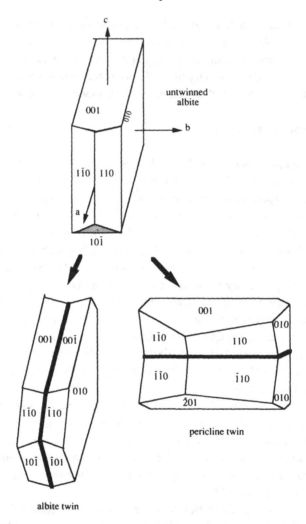

Fig.7.3 Natural faces and crystallographic axes in Na-feldspar. The crystal can develop two types of transformation twins: Albite twins (left) and Pericline (right). The Albite twin wall is parallel to the (010) plane whereas the Pericline twin wall is (nearly) perpendicular to this plane with an orientation which is not directly determined by a crystallographic plane.

given in matrix notation (x,y,z) = (1,2,3) and Voigt notation. The axes are defined with y being the crystallographic b axis and x,z as two Cartesian coordinates in the (010) plane (Redfern et al. 1987, Salje et al. 1985). The condition that the walls are free of strain in equation 7.5 leads to the two solutions y=0 and z= $-(e_6/e_4)x$ (Salje et al.1985a). The first condition represents the Albite twin law with the fixed twin plane (010) (Fig.7.3). This twin plane is a W-wall. The second solution is defined by the unit vector in the y-direction and a second vector which satisfies $z/x = -e_6/e_4$. This twin wall is called the Pericline wall. It has no fixed crystallographic orientation and changes as a function of the ratio of the two strain components e_4 and e_6. This twin plane is a W'-wall. The orientational relationship between the twin walls and the crystallographic axes is shown in Fig.7.3.

Domain orientations predicted by the condition in equation 7.5 represent an approximation because we have not taken into account the other two energy terms of the total wall energy. Furthermore 7.5 describes the orientation of the domain walls in terms of the lattice planes of the high-symmetry phase. Since the spontaneous strain changes the orientation of these lattice planes in the low-symmetry phase, small rigid body rotations of the domains must be incorporated to achieve physical contact between the domains (see experimental observations by Schmid et al. 1989). More striking however, is the effect of inter-wall interactions as discussed by Yamamoto et al. (1977a,b), Torres et al. (1982a,b), Salje et al. (1985a) and Palmer et al. (1988,1989). Following the arguments of Salje et al. (1985a), we can distinguish between three geometrical configurations, namely the wall intersection, the interaction between a right angle domain and a planar domain wall and the formation of needle twins. In order to do this we have first to quantify the stresses which occur in crystal regions close to the domain walls, describing these stresses in terms of dislocation densities.

7.1 Description of domain walls in terms of dislocation densities

One possible and useful way to envisage the energy stored in a domain wall is to consider the twin wall as an internal surface between two crystals, i.e. a so-called 'bicrystal'. Such a wall, in a bicrystal, is shown in the high resolution transmission electron microscope (TEM) image in Fig.7.4. We see that misfits between the two crystals lead to a periodic formation of dislocations in the wall which then create a large strain field around the dislocations. These strain fields, and the associated stress field, appear in the TEM image as dark areas which emanate from the interface. Similar misfits, dislocations and stress fields may also occur in the twin wall and we can formally describe the break of the lattice planes in the twin walls as a continuous distribution of dislocations such that the lattice on either side of the domain wall corresponds

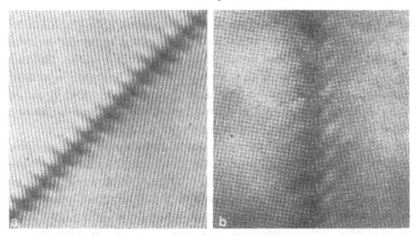

Fig.7.4 High resolution TEM image of a twin wall in a bicrystal (Reimer 1984):
a) 0.204 nm lattice fringes in common [110] orientation,
b) crossed lattice image at a boundary in [010] orientation.

to the other by a simple rotation. The walls in this case are tilt boundaries
with intersections which create incompatibilities of the adjacent lattices. A
large body of published data exists concerning the geometrical consideration
of dislocation densities in such interfaces: the basic ideas which we follow
here were formulated by Kleman and Schlenker (1972) for the investigation
of magnetostrictive properties of ferromagnetic crystals.

The dislocation density in the domain wall is denoted by the tensor α with
the components α_{ik} corresponding to the dislocation direction i and the
direction of the Burgers vector k. The diagonal elements of the tensor α are
hence densities of screw dislocations and the off-diagonal elements represent
edge dislocation densities. The tensor of the dislocation densities is directly
related to the differences between the spontaneous strains of two adjacent
twin domains by:

$$\alpha_{11} = \Delta e_{12} \qquad \alpha_{12} = \Delta e_{22} \qquad \alpha_{13} = \Delta e_{23}$$

$$\alpha_{21} = -\Delta e_{11} \qquad \alpha_{22} = -\Delta e_{22} \qquad \alpha_{23} = \Delta e_{13}$$

$$\alpha_{31} = 0 \qquad \alpha_{32} = 0 \qquad \alpha_{33} = 0 \qquad\qquad 7.8$$

where the z-axis is perpendicular to the twin plane.

Let us illustrate the dislocation densities for the case of Na-feldspar. The
relevant Cartesian system for the Albite twin wall (W-type) with the z axis

perpendicular to the twin plane is: x parallel to c*, y parallel to a and z parallel to b. The corresponding strain tensors are then transformed from equation 7.6 to:

$$e_{ik}(S_1) = -e_{ik}(S_2) = \begin{pmatrix} 0 & 0 & e_{23} \\ 0 & 0 & e_{12} \\ e_{23} & e_{12} & 0 \end{pmatrix} \qquad 7.9$$

The components of the dislocation density tensor are then:

$$\alpha_{13} = 2e_{12} = e_6 \qquad\qquad 7.10$$

$$\alpha_{23} = -2e_{23} = -e_4 \qquad\qquad 7.11$$

All other components are zero. Both non-vanishing components indicate edge dislocations, their sum represents a pure tilt boundary. The lattice on one side of the twin wall is rotated with respect to the other by:

$$\omega = \sqrt{e_4^2 + e_6^2} = e_s \qquad\qquad 7.12$$

The spontaneous strain is hence identical with the angle of rotation between the two twin domains.

The orientation of the Pericline wall (W'-type) is defined by the condition:

$$z = -\frac{e_6}{e_4}x \quad \text{or} \quad z = -x\tan\sigma \qquad\qquad 7.13$$

where σ is the angle of the rhombic section often used in mineralogy.

The corresponding coordinate system with the new z-axis perpendicular to the twin plane is:

$$x_1 = x\cos\sigma - z\sin\sigma$$

$$x_2 = y$$

$$x_3 = x\sin\sigma + z\cos\sigma \qquad\qquad 7.14$$

The transformed tensors of the spontaneous strains are:

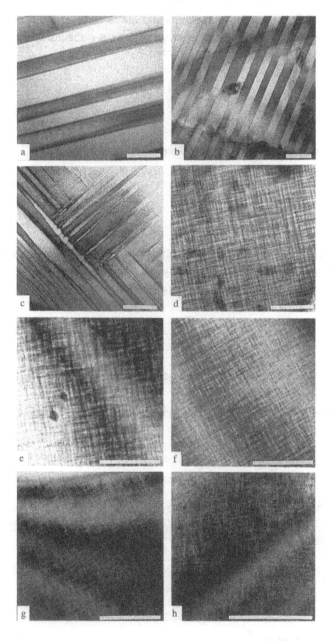

Fig.7.5 Series of domain patterns in superconducting $YBa_2Cu_3O_7$ doped with Co.
Doping leads to a structural phase transition as a function of concentration (under
isothermal conditions) in the same way as using temperature as the external
parameter. The critical concentration C_c is ca. 2.4% Co replacing Cu. The
concentrations in this figure are 0% (a), 1% (b), 2% (c), 2.5% (d), 2.8% (e), 3% (f),

$$S_1 = -S_2 = \begin{pmatrix} 0 & e_{12}\cos\sigma + e_{23}\sin\sigma & 0 \\ e_{12}\cos\sigma + e_{23}\sin\sigma & 0 & -e_{12}\sin\sigma + e_{23}\cos\sigma \\ 0 & -e_{12}\sin\sigma + e_{23}\cos\sigma & 0 \end{pmatrix} \quad 7.15$$

The non-vanishing components of the Burgers vector are:

$$\alpha_{11} = -\alpha_{22} = 2e_{12}\cos\sigma + 2e_{23}\sin\sigma \qquad\qquad 7.16$$

$$\alpha_{23} = -2e_{12}\sin\sigma + 2e_{23}\cos\sigma \qquad\qquad 7.17$$

indicating a mixture between edge dislocations (α_{23}) and screw dislocations ($\alpha_{11} = -\alpha_{22}$). For small values of e_{12}, the screw dislocations disappear and the two adjacent lattices can be transformed into each other by a pure tilt motion. The tilt angle in this limit is identical with that of the Albite twin law.

7.2 The intersection of two domain walls.

a. Experimental observations

Ferroelastic twin domains can penetrate a crystal over its entire diameter without any visible interruption of the domain walls. If several domain walls with parallel orientation occur simultaneously we find homogeneous domain patterns over large areas of the crystal (Fig.1.3). This homogeneous pattern is, however, seriously disturbed if domains with different orientations occur simultaneously in the same part of the crystal. Figure 7.5 shows a series of domain patterns in superconducting $YBa_2Cu_3O_7$ doped with Co. The intersection of two domains at approximately right angles, leads to large strain fields which emerge from the intersection (Fig.7.5c). These strain fields give rise to strain contrast in transmission electron microscopy (TEM) and

Fig.7.5 (cont'd) 5% (g), 7% (h). The scale bar is $0.1\mu m$. The low symmetry phase shows irregular (a) and regular (b) striped twin patterns. Close to the transition point, patches of different twin orientations occur (c). The deformation of the twin walls near the intersections are discussed in detail in Chapters 7.2-7.4. In the high symmetry phase we observe a tweed pattern (d-h) with weak strain contrast. These patterns are discussed in Chapter 8 (courtesy A Putnis, Cambridge).

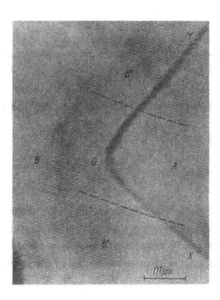

Fig. 7.6 Right angle domain walls in $Gd_2(MoO_4)_3$. (After Yamamoto et al. 1977).

become visible as a wedge-shaped darker area, with the wedge pointing towards the joint of the two twin domains. Under higher resolution TEM, the fine structure of the domain shape near the junction becomes apparent (Fig.7.6). It appears that the twin walls bend continuously from one wall into another with rounded tips between them. We shall see that this avoidance of a pointed intersection of the two domain walls leads to a considerable reduction of the strain energy from their intersection. It will also be shown that the curvature of the rounded tip is directly proportional to the local stress field which is directly determined by the magnitude of the spontaneous strain and the local elastic constants of the material: the radius of the rounded tip is therefore proportional to the spontaneous strain and a suitable combination of elastic constants so that mechanically hard materials show systematically larger radii than soft materials.

b. Theoretical considerations

The strain fields related to the domain boundaries may increase near the intersection of two boundaries and some general rules for the shape of the intersecting walls can be derived using the general condition that the walls

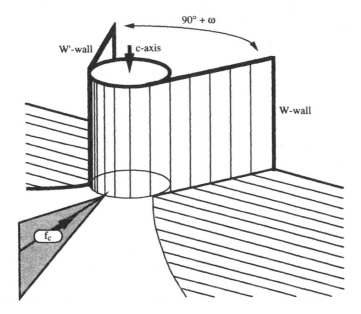

Fig.7.7 Intersection between a W wall and a W' wall. The junction is parallel to the c-axis. The geometrical configuration of the junction is related to a continuous bending of the W' wall around a cylinder around the c-axis into the W wall. The angle between the walls is $90° \pm \omega$. The force which is needed to bend the tip of the junction is f_c (equation 7.2.3).

will change until the wall energy reaches a minimum. Let us first consider the tilt components of the dislocation densities alone. The strain due to the intersection of a W and a W' wall is in this case identical to an inclination located at the junction of the walls with a rotation axis in the junction and a rotation angle equal to the sum of the tilt angles, ω, of the W and W' wall. The actual angle between the two walls is hence not exactly $90°$, as it would be if the lattice matrix of the high-symmetry phase were preserved. As the lattice is sheared in the low-symmetry phase, the effect of the spontaneous strain has to be taken into account. The additional tilt angle, ω, is numerically identical with the spontaneous strain for pure tilt boundaries and we find that the actual angle between the intersecting twin walls is $90° + \omega$ or $90° - \omega$ depending on the orientation of the tilt. The geometrical configuration of the intersection between the W and W' wall is depicted in Fig.7.7.

Taking a circular path around the junction thus leads to a periodicity of $360° \pm \omega$, i.e. contains a misfit of $\pm \omega$. The surrounding lattice has now to be distorted in order to compensate for this misfit. An illustrative way of calculating the necessary stress field stems from Kleman and Schlenke (1972) who have argued that the relevant stress field approximates to that of an edge dislocation. Yamamoto et al. (1977a,b) have subsequently calculated the stress field in the isotropic approximation and found:

$$\sigma_{11} = \frac{\mu\omega}{2(1\text{-}\upsilon)}\left(\ln\frac{\rho}{R} + \frac{y^2}{\rho^2}\right) \qquad\qquad \sigma_{12} = -\frac{\mu\omega}{2(1\text{-}\upsilon)}\frac{xy}{\rho^2}$$

$$\sigma_{22} = \frac{\mu\omega}{2(1\text{-}\upsilon)}\left(\ln\frac{\rho}{R} + \frac{x^2}{\rho^2}\right) \qquad\qquad \sigma_{23} = \sigma_{13} = 0$$

$$\sigma_{33} = \frac{\mu\omega}{2(1\text{-}\upsilon)}\left(\ln\frac{\rho}{R} + \frac{1}{2}\right)$$

$$7.18$$

The parameters are: the Lame' coefficient μ, the Poisson's ratio υ, the distance from the junction ρ and the characteristic length scale of the strain field R. The coordinate system is chosen in such a way that the z axis points along the junction and the x and y axes are in the two intersecting planes (ignoring the geometrical effect of ω).

We can now approximate this stress field using a simple geometrical construction which is independent of the assumption of actual dislocations in the wall. We have seen that the essential effect of the wall intersection is the fact that the sum of the angles between the domain wall going round the junction is either 360° + ω or 360° - ω but never 360° as it should be in an unstressed crystal. There is hence always a misfit of the lattices present which leads to a wedge in the crystal, which sees a superposition of material (+ sign) or is empty (- sign). In order to make the crystal continuous again we have to squeeze material in or out of this wedge. As the stress field acts effectively as a shear stress with a rotation axis in the junction, we can imagine the effect of this field as produced by a single force f_c directed towards the junction and rotated by 45° relative to both the x and y axis (Fig.7.8). It is fortunate for the theorist that this result is identical for a dislocation model as well as a simple deformation model so that it does not matter if we have or have not actual dislocation lines in the twin wall or if, as we shall assume in Chapter 8, the lattice is flexible enough to compensate for all misfits by elastic deformations. All theoretical results in this chapter are hence correct in both cases.

Throughout this chapter we have assumed that the shape of the domain wall is entirely dominated by the elastic properties of the surrounding lattice and not by local structural properties. In this case, continuum theory is applicable

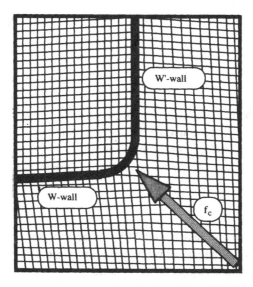

Fig.7.8 Walls, lattice planes and force field f_c projected along the c-axis of the junction (Fig.7.7) The walls relate to twinning at the top and the left hand corner of the figure. At the junction, the walls bend into each other with a rounding of the tip. Lattice planes are highly distorted close to the junction. These distortions and the rounding of the walls is related to the force f_c via equation 7.2.3.

and it therefore follows that the stress field has to satisfy the classical potential equation which is, in cylindrical coordinates:

$$\Delta\sigma_r = \frac{1}{r}\frac{\partial\sigma_r}{\partial r} + \frac{1}{r^2}\frac{\partial^2\sigma_r}{\partial\varphi^2} + \frac{\partial^2\sigma_r}{\partial r^2} = 0$$

7.19

where σ_r is the radial component of the stress field, r is the distance from the junction and φ is the angle around the junction. The zero point is again in the junction which is parallel to the z-axis. The angle is defined counterclockwise from the extrapolated vector of the force f_c (Fig.7.7 and 7.8). The solution under the boundary condition that any tangential component of the stress field has to vanish (Fig. 7.7) is:

$$\sigma_r \approx \frac{-2 f_c}{2\alpha + \sin 2\alpha}\frac{\cos\varphi}{r}$$

7.20

which represents a cylinder with its axis on the intersection of the two domain walls. The angle α is the angle between the two planes and can be approximated as $\pi/2$ so that the final solution is in good approximation for values of spontaneous strain of some percent:

$$\sigma_r \approx -\frac{2 f_c}{\pi} \frac{\cos \varphi}{r} \qquad\qquad 7.21$$

The shape of the domain boundary follows now from the condition that the stress inside the wedge must be compensated by other forces such as represented by the surface energy in equation 7.21. Let us summarise the compensating stress field as σ_0 and write the equilibrium condition as:

$$\sigma_r = \sigma_0 \qquad\qquad 7.22$$

The shape of the domain boundaries near the junction follows the geometrical shape of the equilibrium stress, therefore, and we find that the stable configuration is a rounded tip of the intersection of the twin walls. The radius of the rounded tip is r_c with:

$$r_c = -\frac{2 f_c}{\pi \sigma_0} \qquad\qquad 7.23$$

The force f_c was introduced as a measure for the stress field close to the junction (equation 7.18) which is, in the isotropic approximation, proportional to the spontaneous strain and the elastic stiffness. Anisotropic fields lead to essentially similar results if the Lame coefficient and the other elastic parameters are replaced by the appropriate combination of elastic constants. This leads us to the conclusion that the curvature of the rounded tip between two intersecting twin walls increases with an increase of the spontaneous strain (e.g. by lowering the temperature $T \ll T_c$) and is larger for stiff materials than for soft materials.

Main result: the intersection between two orthogonal twin walls is rounded with a radius which depends on the elastic stiffness of the crystal.

7.3 The S-domains and triple junctions

a.Experimental observation

A domain pattern that is commonly observed in the TEM is a twin pattern wiggle in the domain boundary,mostly in the form of an S-shape.

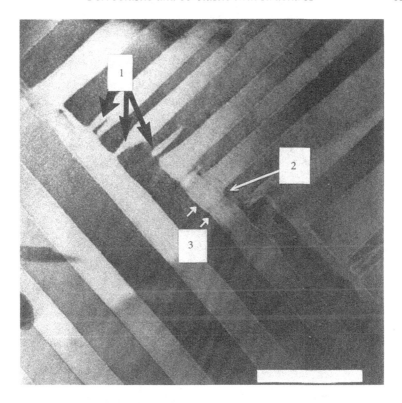

Fig.7.9 Domain structure of superconducting $YBa_2Cu_3O_7$: 1% Co (scale bar = $0.1\mu m$). Two striped domain patterns interact. Close to the intersection, needle shaped domains develop (1). The intersections between two walls are rounded (2). Non-intersecting walls close to such intersections show characteristic modulations.

A typical example for a ferroelastic $YBa_2Cu_3O_7$ superconductor is shown in Fig.7.9. In feldspars and other co-elastic minerals with twin structures, one often finds domain patterns which are similar to those in Fig.7.10.

None of these wall deformations occur in isolated twin domains, however, and it appears that it is the interaction between different twin walls which leads to the bending of the twin walls. Wall deformations of the S-type occur, in most cases, close to right angle intersections, as discussed in the last chapter. The distance from the intersection can be large, however, and it might not always be obvious from inspection of high resolution TEM images that these interboundary effects exist because the intersection might be outside the imaged area of the crystal. Yamamoto et al. (1977b) have shown that wall bending can occur in oxides over distances as large as $1\mu m$ between

1 µm

Fig.7.10 Twin structure close to an irregular interface in alkali feldspars from a syeno-gabbro (klokken intrusion). Cryptoperthite is seen at the left with needle or trumpet-shaped W or W' lamellae. They do not reach the interface which is locally perpendicular to the twin walls and bends into steps (Courtesy of W.L. Brown).

the domains. Triple junctions are domain patterns in which three twin domains interact under the formation of two junctions. A typical example for lead vanadate is shown in Fig.7.11.

The appearance in the TEM image is again similar to an S-shape domain with the essential difference that two right angle domains intersect with the planar domain wall at the points of its maximum distortion. The associated stress fields are normally greater than in the proper S-shape domains and might lead to a significant loss of contrast in the TEM image. If several of these triple junctions occur close to each other, a tweed like-structure occurs, similar to the microstructure often seen in feldspars (Fig.7.12).

Fig.7.11 Triple junction in $Pb_3(VO_4)_2$ (Courtesy of Van Tendeloo and Van Landuyt).

Fig.7.12 Photomicrograph of two examples of microtwinned plagioclase with W' (Pericline) walls in horizontal and W (albite) walls in the vertical orientation. Blocks of Albite and Pericline twinned materials form a coarse tweed pattern. Note that only very few intersections (i.e. triple junctions) occur. Block boundaries are distorted and deviate from the (010) direction of the lattice. Scale bars are 0.1μm. (Courtesy of W.L. Brown, Nancy).

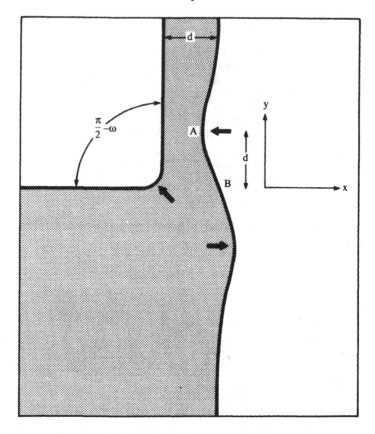

Fig.7.13 The forces acting on a non-intersecting wall (right hand side) close to a junction between two twin walls. The non-intersecting wall is bent into an s-shaped wiggle.

b. Theory

Domain walls parallel to one side of a right angle domain wall interact with the stress field induced by the edge dislocation near the rounded corner. The stress field in equation 7.19. causes a force in each dislocation in the planar wall. The force per unit area is given by the stress field multiplied by the Burgers vector:

$$f_x(xy) = -\sigma_{xy}(xy)\, b_x = -\frac{\mu\omega^2}{4\pi(1-\upsilon)}\frac{xy}{x^2+y^2} \qquad 7.24$$

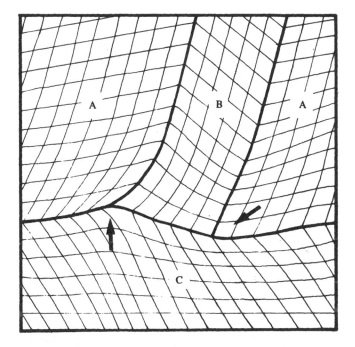

Fig.7.14 Lattice distortion around two twin walls of the one type (e.g. W, shown as vertical) interacting with a wall of the second type (e.g. W', shown as horizontal).

where b_x is the x-component of the Burgers vector. The force acts perpendicular to the wall at a distance x=d. We are again using the isotropic approximation which suffices to illustrate the physical mechanism. In this approximation, the maximum force occurs on the line x=y which intersects the planar domain boundary at y=d (Fig.7.13). The twin wall shows its largest distortion at this point with an attractive force pulling the wall towards the intersection at positive y values and pushing it away at negative y values. The dislocations at y=0 are basically unaffected and this point remains undistorted. The inter-domain interaction tends, therefore, to distort the planar domain wall into an S-shape as shown in Fig.7.9 and 7.10.

The distortive force is compensated by other forces which are mainly related to the increase of surface energy. This increase may, for example, create new dislocations in the domain wall which will again increase the surface energy. A stable domain configuration is reached when the energy gain from bending the planar domain wall into an S-shape is compensated by the energy loss of the surface energy and other energies related to restoring forces. The equivalent force fields in the case of triple junctions are a superposition of two parts:

Fig.7.15 Optical micrograph showing needle domains in lead phosphate.

a that acting on the domain wall parallel to the vertical y axis, called $f_x(y)$ and $f_y(y)$ and
b those acting along the x axis with $f_x(x)$ and $f_y(x)$. Their functional forms have been calculated by Torres et al. (1982a):

$$f_x(xy) = \sigma_{xy}(xy) b_x = - \frac{\mu\omega^2}{4\pi(1-\upsilon)} \frac{xy}{x^2 + y^2} \qquad\qquad 7.25$$

$$f_y(xy) = - \sigma_{xy}(xy) b_y = - \frac{\mu\omega^2}{4\pi(1-\upsilon)} \left(\frac{1}{2} \ln \frac{x^2 + y^2}{R} \frac{y^2}{x^2 + y^2} \right) \qquad\qquad 7.26$$

The equivalent bending of the lattice planes is illustrated in Fig.7.14. The elastic distortions compensate for the angular misfits between the junctions a-c-b and b-c-a, the arrows indicate the dominant direction of the effective forces which lead to the distortion of the twin planes.

Main result: a non-intersecting twin wall close to a wall intersection is bent into an s-shape with a profile given by the equations 7.25 and 7.26.

Fig.7.16 Group of intersecting domains in a matrix with a regular twin pattern (striped phase). The domains sharpen up to needles avoiding actual intersections. The regular parallel domains on the right hand side terminate in a comb of needle domains in front of the first orthogonal wall. (scale 0.1μm, $YBa_2Cu_3O_7$, courtesy of A. Putnis, Cambridge).

7.4 The formation of needle shaped domains

a. Experimental observations

The best known feature of ferroelastic domains which is also a common feature in twinned co-elastic materials is the needle domain. The size of these needles varies between some millimeters and some ten Ångstroms. In

Fig.7.15 we see a typical ferroelastic crystal under crossed polarizers. The matrix of this crystal is one large single-domain. Inside this large domain are small needle shaped domains. The domain walls are straight with a sharp tip at the end. Several of these needle domains can occur in this crystal. It is normally easy to extend or contract these needles along their axes by external stresses (Fig.7.15 and 7.16).

The origin of these needles becomes clear by observation under the electron microscope. In Fig.7.17 we see the typical appearance of such needle tips close to an intersection with another domain wall which is approximately perpendicular to the needle axis. In the beginning of the observation two right angle domains may form a triple junction. When this part of the crystal is heated by the electron beam the two neighbouring junctions move together and eventually merge to one intersection of three domain walls. Further heating leads to the retraction of the tip of the needle from the straight domain wall. This pulling back of the needle occurs rapidly and can easily be observed under the optical microscope. The needle domain will then completely disappear if there is no defect in the crystal to act as a pinning centre. Repeated heating can unpin the domain wall and eventually lead to a domain-free crystal. The early stages of contraction of the two junctions, their merger and the final retraction of the needle can be see simultaneously in Fig.7.17. When the needles exceed a critical thickness, one finds the splitting of the end into two or more needle tips (Fig.7.17).

b. Theory

The theoretical approach used for understanding needle domains is equivalent to that used for the other domain structures discussed so far. We assume that the elastic energy on a mesoscopic scale has to be minimised where the strain fields due to wall junctions are the dominant source of energy. Following the argument used for the explanation of triple junctions, we can simplify equations 7.25 and 7.26 for two junctions with the same position on the vertical y axis. The total force is then in the elastic approximation:

$$F(r) = - \frac{\mu\omega^2}{2\pi(1-\upsilon)} r \ln \frac{R}{r}$$

7.27

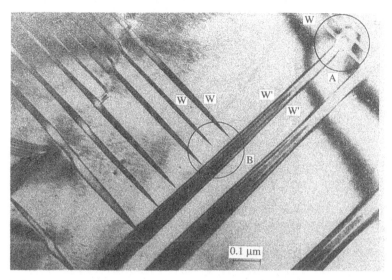

Fig.7.17a) Group of needle shaped domains in $Pb_3(PO_4)_2$. The domains near the lower left hand corner are still attached to the W' wall whereas those within the circle denoted B are already separate. A shows triple junctions. (Courtesy of J. Torres).

Fig.7.17b) Needle domains in the high T_C superconductor $YBa_2Cu_3O_7$(Courtesy of Van Tendeloo and Van Landuyt)

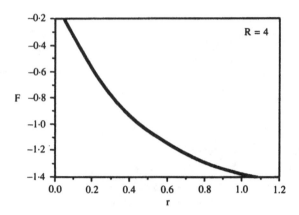

Fig.7.18 Attractive force between two junctions (equation 7.4.1) for R=4 in relative units.

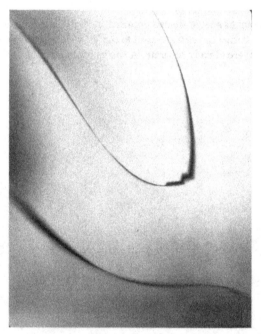

Fig.7.19 Anti-phase boundary (with respect to cation ordering). Note the straight sections in the otherwise rounded domain boundary. (Courtesy G. L. Nord Jr)

This force decreases as a function of the distance between the junctions (Fig.7.18) leading to a wall configuration shown in Fig.7.19. The restoring force is again due to the surface energy because the formation of the needle tip leads to an increase of the wall area. Once the two junctions have merged, there is no further attractive force between them. The surface energy, however, can now be reduced dramatically when the needle migrates back into the matrix and finally disappears completely.

Main result: two junctions with a common twin wall attract each other and merge to form a needle domain.

We remark here, that anti-phase boundaries (APB) (Amelinckx and Van Landuyt 1976) can be treated in a similar way as twins, although their strain coupling is often smaller. A typical example for a system with weak strain coupling is shown in Fig 17.19 where round APBs show straight, twin-like sections when the orientation is close to the strain-related lattice planes (equivalent to W and W' walls with junctions).

8

DOMAIN MOBILITIES AND ELASTIC INSTABILITIES IN FERROELASTIC AND CO-ELASTIC MATERIALS

The close connection between twin walls and kinks (or solitary waves) is discussed using a local potential of the Landau-Ginzburg type. Moving kinks define simple kinetic rate laws. Some algebra is necessary to derive expressions for the wall thickness and the wall energy in a second order and a tricritical phase transition. If the reader is more interested in the final results summarised in 8.3 than in their derivation and examples, he or she can ignore 8.2. In 8.4 the correlation between acoustic instabilities and twin wall formation is demonstrated using Al,Si disordered Na-feldspar as an example.

8.1 Solitary waves and a simple rate law

We have seen that domain structures can change rapidly when the thermodynamic condition of the sample, e.g. its temperature, changes even slightly. One example was the rapid formation of needle twins and their annealing. A second example is the lateral movement which can lead to the formation of periodic domain patterns. Snap-shots of such motions are shown in Fig.8.1 - 8.3.

In this chapter, the main physical features of these motions are described for a continuous, defect-free lattice. We shall discuss the strain profiles, the thickness and the energy of twin walls within the framework of the physical behaviour of solitary waves.

The starting point is the observation that the main characteristic feature of a co-elastic phase transition is the role played by the elastic interactions which act over long interatomic distances. Let us imagine a situation in the crystal where the phase transition is imminent but where only small pertubations of the crystal structure probe the different structural instabilities. Such fluctuations will occur locally and cannot be described in the homogeneous approximation of the Landau potential as in equation 3.6. We now define an order parameter $Q(r)$ on a mesoscopic scale over which there are enough atoms present so that this order parameter has, at least approximately, a

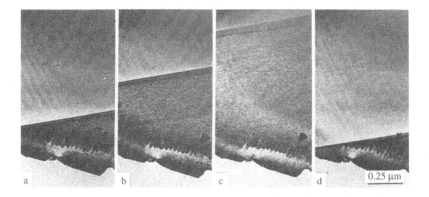

Fig.8.1 Moving wall (a→d) between two ferroelastic domains in $Pb_3(VO_4)_2$ (Courtesy of Van Tendeloo and Van Landyut).

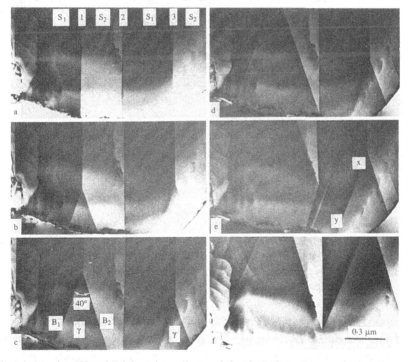

Fig.8.2 Moving W and W' domain walls (a→f) in $Pb_3(VO_4)_2$. The vertical walls 1,2,3 separate the ferroelastic domains S_1 and S_2. In Fig.8.2b new domains γ of the paraelastic phase move along the ferroelastic twin walls 1 and 2. This forms new walls B1 and B2 between the high symmetry and the low symmetry domains. The phase γ ultimately (c,f) dominates this part of the crystal which is then paraelastic (Courtesy of Van Tendeloo and Van Landyut).

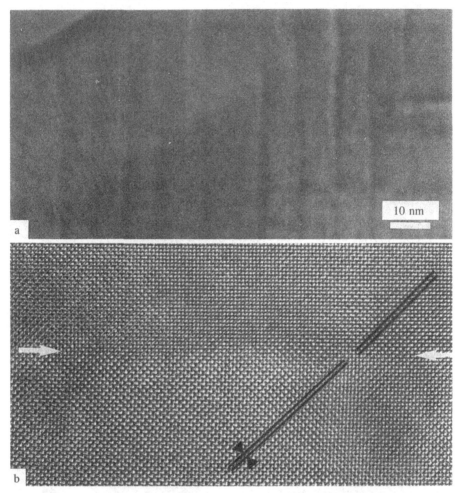

Fig.8.3 Periodic array of domain walls in the High T_C Superconductor $YBa_2Cu_3O_7$ (a) and high resolution image of a wall (b). The domain walls appear as diffuse interfaces which is due to rather large wall thicknesses (see also Fig.8.6). (Courtesy of Van Tendeloo and Van Landyut).

thermodynamic meaning. The length scale is, on the other hand, small enough to 'see' the elastic fluctuations. The space coordinate, r, is hence measured over several tens of Ångstroms and we can formulate the effect on the Gibbs free energy of inhomogeneous fluctuations of $Q(r)$. A rather elegant way of treating this problem is related to the evaluation of the Gibbs free energy introduced in Chapter 3 (equation 3.1) via a partition function:

$$G = -k_B T \ln \sum_N e^{-\mu N/k_B T} \int e^{\frac{-E}{k_B T}} \, d\Gamma$$

8.1a

where μ is the chemical potential, Γ is the total set of variables which might change during the phase transition (i.e. the whole phase space, in the language of thermodynamics) and E is the relevant energy change leading to the phase transition. If the parameters in Γ are highly correlated, they can be expressed by all Fourier components of the order parameter. Furthermore, we are only interested in the Q-dependence of G so that the direct effect of the chemical potential μ is irrelevant here. Our first objective is then to find the energy function E which leads to the correct Gibbs free energy. This energy function is called the effective Hamiltonian H_{eff} and equation 8.1a can be rewritten using this term as:

$$G(Q) = -k_B \, T \ln \int_{\Gamma} \exp\left(- H_{eff}/k_B T\right) V \prod_k dQ_k$$

8.1b

where the integration extends over all Fourier components of the order parameter $Q(r)$. As the length scale over which the order parameter varies is long in co-elastic materials, we can safely assume that there is a cut-off wavevector above which there are no significant contributions to the integration in equation 8.1. The symmetry of the order parameter on the mesocopic scale in a co-elastic material is basically the same as its thermodynamic expectation values, with the role of the fluctuations limited to 'warping' the elastic strain fields. This means that the phase volume $\{Q_k\}$ is restricted to values close to the expectation value Q_k which leads to the conclusion that the role of the Landau potential is now taken over by the effective Hamiltonian. We can then write H_{eff} as a function of the mesoscopic order parameter in the same way as we did for the macroscopic Landau potential. Since $Q(r)$ varies slowly, we include in the expansion only the terms of lowest order in its derivatives. The form of the effective Hamiltonian is then:

$$H_{eff} = \int \left\{ \frac{1}{2} a \, Q^2 + \frac{1}{4} b \, Q^4 + \frac{1}{6} c \, Q^6 - hQ + \frac{1}{2} g' (\nabla Q)^2 \right\} d\,V$$

8.2

For a limited phase volume, the partition function (8.1) can be solved and we find the **Landau-Ginzburg** expression of the Gibbs free energy:

$$G = \frac{1}{2} A(T\text{-}T_c) \, Q^2 + \frac{1}{4} BQ^4 + \frac{1}{6} CQ^6 - H \, Q$$

$$+ \frac{1}{2} g \left(\nabla Q\right)^2 = L(Q) + G_{Ginzburg}$$

8.3

where Q(r) for vanishing gradient terms is again the homogeneous, macroscopic order parameter Q and G becomes the Landau potential.

The arguments which lead us to the Landau-Ginzburg energy imply the restriction of the phase volume $\{Q_k\}$ to only 'harmless' modulations of the order parameter. This assumption does not apply if $\{Q_k\}$ has to be increased; for instance in magnetic phase transitions or for all phase transitions of the order-disorder type (e.g. Zoo and Liu 1976, Giddy et al. 1989, Tang et al. 1986). This is independent of whether or not a mean-field approximation is applied but relates to the essential assumption that entropy fluctuations contribute to the effective Hamiltonian in Landau-Ginzburg theory only in the lowest order ($S \propto Q^2$). On the other hand, large phase volumes always imply a highly non-harmonic entropy even for the most simple Bragg-Williams approximation (e.g. Salje 1988a). Although the rather stringent assumption of harmonic entropy in Landau-Ginzburg theory might appear excessive from a purely theoretical point of view, the experimental evidence in ferroelastic and co-elastic phase transitions is in general agreement with this idea and we shall use Landau-Ginzburg theory in what follows.

Let us now envisage the moving domain walls as indicating a local energy variation. The driving force of the movement is the tendency of the system to lower its total energy via a kinetic relaxation towards equilibrium. On a macroscopic level, the order parameter Q(r) will then follow a kinetic rate law which was first introduced by Landau and Khalatnikov (1954) :

$$\frac{\partial Q}{\partial t} = -\frac{1}{\tau}\frac{\partial G}{\partial Q} \qquad\qquad 8.4$$

where τ is a time constant. This rate law is normally referred to as the **time-dependent Landau-Ginzburg or Ginzburg-Landau equation.** We shall see more general rate laws in Chapter 12; the simple Landau-Ginzburg approach suffices for the present discussion.

Let us now consider a propagating domain wall. The profile of the interface will preserve its shape during the propagation because of the two competing forces which drive the motion: firstly, the Landau potential in G pushes the order parameter Q into one of its stable minima (e.g. Q=0 and $Q_1, Q_2 \neq 0$ for the domains 1,2). The second force is due to the gradient energy and counteracts the Landau potential. Assuming that there is no orientational dependence of the order parameter, we can write the local equilibrium condition between the gradient 'force' $g\{\partial^2/\partial r^2 + (d-1)/r \; \partial/\partial r\}Q$, the 'force' due to the Landau potential $\partial/\partial Q \; L(Q)$ and a frictional 'force' $1/\tau \; \partial Q/\partial t$ as:

$$g\left\{\frac{\partial^2}{\partial r^2} + \frac{d-1}{r}\frac{\partial}{\partial r}\right\}Q - \tau\frac{\partial Q}{\partial t} = \frac{d}{dQ}L(Q)$$

8.5

where L is the Landau potential. The parameter d indicates the dimension of the system (d=1,2,3). It is now easy to show that this equation of motion does not change in the one-dimensional case if we replace Q(r,t) by Q(r-R(t)) so that the interface will indeed not change during the motion. In two and three dimensions we find that this invariance is no longer strictly correct, although it still applies to a good approximation if the thickness of the wall is small compared with the size of the homogeneous parts of the crystal.

The actual shape and the energy of the wall itself is then determined by 'local' forces such as Newtons acceleration force $\partial^2 Q/\partial t^2$ and the forces derived from the Landau-Ginzburg potential $\partial G/\partial Q$ for long interaction lengths:

$$\left\{c_0^2\frac{\partial^2}{\partial r^2} - \frac{\partial^2}{\partial t^2}\right\}Q(r,t) = A(T-T_c)Q(r,t) + B\,Q^3(r,t) + ..$$

8.6

where c_0 defines some characteristic velocity. Equation 8.6 is, in fact, the most simple Euler-Lagrangian equation of a solitary wavewhich we now discuss before we return to the macroscopic rate law in eq. 8.5. Equation 8.6 is, as eq.8.5, invariant with respect to a shift of the space coordinate and can be simplified if rewritten in a new coordinate ξ which moves with the solitary wave:

$$\xi = \frac{r-vt}{\sqrt{1-\frac{v^2}{c_0^2}}}$$

where v is the velocity of the wave. The profile of the moving domain wall in eq. 8.6 has then the well known solution (e.g. Collins et al. 1979):

$$Q(r,t) = Q_0\tanh\frac{r-vt}{w\left(1-\frac{v^2}{c_0^2}\right)^{\frac{1}{2}}}$$

8.7

where 2w is the wall thickness at zero velocity:

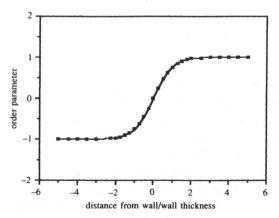

Fig.8.4 Shape of the wall between a domain with the order parameter -Q_0 (left) and a second domain with the order parameter +Q_0 (equation 8.7). The units of the order parameter are the equilibrium values (±Q_0), the space coordinate is measured in units of the wall thickness w (equation 8.8).

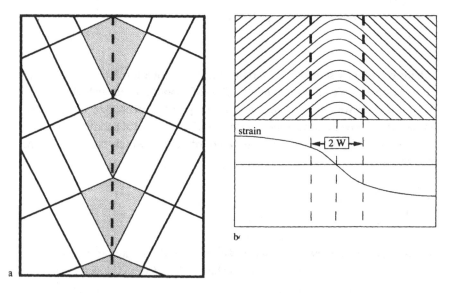

Fig.8.5a) Simple picture of an unrelaxed twin wall (grey area) between two ferroelastic domains. The wall has high elastic energy which often leads to a continuous lattice distortion as depicted in b).

b) Twin wall (top) and correlated spontaneous strain (bottom) in bilinear coupling. The lattice planes are bent continuously in the wall (indicated by the two dashed lines at either end). The equivalent spontaneous strain moves also continuously from its positive equilibrium value e_0 to its negative equilibrium value -e_0. The profile is described by equation 8.7 (see Fig.8.4).

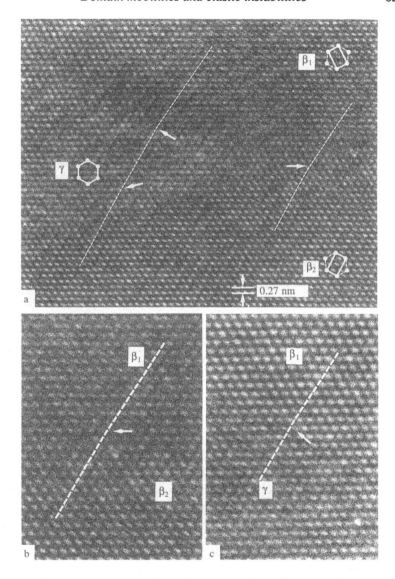

Fig.8.5c) High resolution images along the c-zone in $Pb_3(VO_4)_2$. A needle shaped domain of the paraelastic phase (γ,left hand side) wedges between two domains of the low temperature phase (β_1,β_2 right hand side). The actual domain walls appear as dark lines. Two lattice planes are highlighted by the dotted lines. Their intersection with the twin walls is indicated by arrows. Note the bending of the lattice planes in the twin walls. The lower two images show a magnified part of the wall between the para- and ferroelastic phase (left) and a W-wall between two ferroelastic domains. The lattice planes are bent by 6°, the width of the walls is some 20Å (after Manolikas et al. 1986, courtesy of G. Van Tendeloo).

$$w = \frac{\sqrt{2}c_0}{\sqrt{A|T\text{-}T_c|}}$$

8.8

which diverges as the temperature approaches the phase transition point. The shape of the interface for v=0 is shown in Fig.8.4.

Interfaces of this type will occur between two twin domains or clusters, e.g. with the internal order parameters Q_0 and $-Q_0$ (representing different twin orientations as in Fig.8.5). When the crystal undergoes the phase transition we expect, according to equation 8.7, that these walls will become mobile and increase their thickness simultaneously. Figure 8.6 shows two pinned twin walls in a high T_C superconductor $YBa_2Cu_3O_7$ on either side of a needle domain. The thickness of the walls is ca. 6 unit cells (i.e. ca. 30Å). In terms of a direct observation using TEM, ignoring all effects induced by the preparation of the sample, we would see that all twin walls become 'blurred' while they start to travel (Fig.8.3).

These domain walls are known as **kinks, solitary waves,** or (somewhat inappropriately) **solitons.** Their propagation, with velocity v is accompanied by atomic motions with large amplitudes similar to the switching of the domain structures from one ferroelastic state to another. These atomic motions are essentially nonharmonic and cannot, therefore, be described in a phonon picture of motion. By integration of the Ginzburg energy it can be shown that the velocity is related to the energy required to create a single wall via:

$$E = \frac{E_0}{\left(1-v^2/c_0^2\right)^{1/2}}$$

8.9

where the energy of the static wall E_0 can be understood as follows. Let 2w be the total thickness of the wall. The energy stored in the wall is then given by the mean value of the energy of a particle in the wall, namely $(A(T\text{-}T_c))^2/2B$ in the Landau potential, multiplied by the wall thickness $2w = 2\sqrt{2}\,c_0\,|A(T\text{-}T_c)|^{-1/2}$ giving:

$$E_0 = \sqrt{2}\,\frac{c_0}{B}\,|A(T\text{-}T_c)|^{3/2}$$

8.10

This energy disappears at the transition point when no energy is required to form a twin wall. Lowering the temperature, we find that the wall energy increases with the exponent 3/2. At temperatures slightly below T_C we, thus, find that the energy required to form twin walls is still very low and highly twinned crystals are expected if a crystal (without external stress-fields) is annealed under these conditions. If we now envisage E as the activation energy which has to be overcome in order to create a twin wall, we see

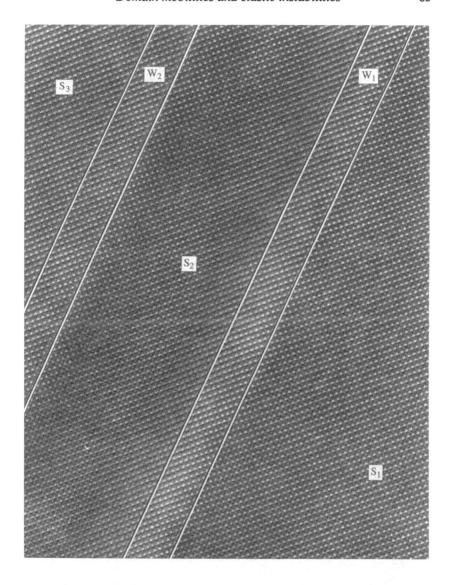

Fig.8.6a) High resolution TEM picture of a twin wall in $YBa_2Cu_3O_7$: 2% Co. Two parallel walls of a needle domain are shown. The shear of the lattice is 1.5°, the wall thickness is ca. 6 unit cells, the diameter of the needle domain (i.e. the distance between the walls) is ca. 12 unit cells. It is often observed that the smallest domain diameter is twice the equivalent wall thickness.

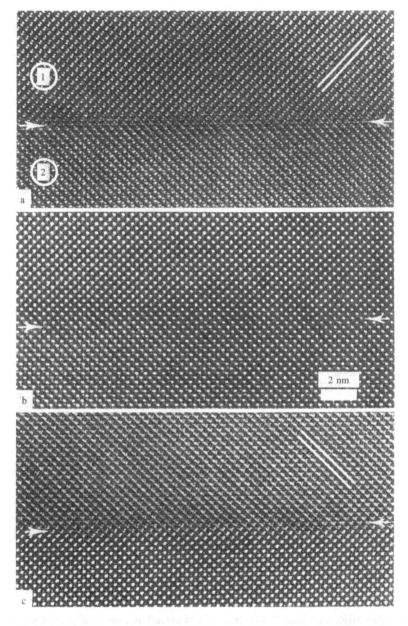

Fig.8.6b) High Resolution Electron Microscope images of a domain wall in
$YBa_2Cu_3O_7$ under different defocus values showing the horizontal domain wall
(Courtesy of Van Tendeloo and Van Landuyt).

immediately that the number of twin walls will depend exponentially on E and, thus, depend on temperature in a second order phase transition according to the relationship:

$$n \propto \exp - \left(a \frac{|T-T_c|^{\frac{3}{2}}}{T} \right)$$

8.11

We expect, therefore, that the wall density under equilibrium conditions drops rapidly for $T \gg T_c$ or $T \ll T_c$. This rapid change of the wall density with temperature is in marked contrast with the linear variation of the phonon energy with temperature or the square-root dependence of the order parameter on temperature. The actual transition point in a ferroelastic material is, thus, often clearly marked by an anomaly of the microstructure and all physical properties related to the wall density may show the same sharp divergency at T_c as the wall density itself.

We now return to the discussion of the rate law in eq.8.5 as applied to the growth and dissolution of domains in a first order phase transition. Assuming that the walls can again be described as solitary waves we transform eq. 8.5 into coordinates of the moving observer who stays at a fixed position X on the interface. The transformed region of the crystal is behind the moving front with X=r-R(t) where R now indicates the thickness of the transformed area. The untransformed part of the crystal is in front of the interface and will be transformed once the interface has travelled through. The time dependence of the order parameter and the size of the transformed region can then be approximated from eq. 8.5 as:

$$g \frac{\partial^2 Q}{\partial X^2} + \left(\frac{1}{\tau} \frac{\partial R}{\partial t} + g (d-1) \frac{1}{R} \right) \frac{\partial Q}{\partial X} - \frac{dL}{dQ} = 0$$

8.12

under the assumption that the wall thickness w is small compared with the thickness of the transformed area R. Note that this equation contains not only the conditions for solitary waves but also the rate law of R if, and only if, the dimension of the system is higher than one. A solution can be found for the condition:

$$\frac{dR}{dt} = v' - g(d-1) \frac{\tau}{R}$$

8.13

which can be substituted into equation 8.12 leading to the steady-state kinetic equation which defines again solitary waves with

Salje

$$g\frac{\partial^2 Q}{\partial X^2} + \frac{v'}{\tau}\frac{\partial Q}{\partial X} - \frac{dL}{dQ} = 0 \qquad\qquad 8.14$$

For a Landau potential L of the type:

$$L(Q) = \frac{1}{2}A(T-T_c)Q^2 + \frac{1}{3}BQ^3 + \frac{1}{4}CQ^4 \qquad\qquad 8.15$$

the relevant solution of eq. 8.14 is a kink with the profile

$$Q(X) = Q_o \, / \left[1 + \exp\left(Q_o\sqrt{\frac{1}{2}\frac{C}{g}}X\right)\right] \qquad\qquad 8.16$$

where Q_o is the equilibrium solution of the low-symmetry phase:

$$Q_o = \frac{3}{4}Q(T_{trans})\left[1 + \sqrt{\left\{1 - \frac{8}{9}(T-T_c)/(T_{trans}-T_c)\right\}}\right]^{\frac{1}{2}} \qquad\qquad 8.17$$

The parameter $Q(T_{trans})$ is the jump of the order parameter at the transition temperature T_{trans}. The shape of the interface is depicted in Fig.8.7. The parameter v' in equation 8.13 is then:

$$v' = \frac{3}{2}\tau\sqrt{\frac{I}{8}}\,g\,A|T_o-T_c|\left[\sqrt{1 + 8\frac{T_o-T}{T_o-T_c}} - 1\right] \qquad\qquad 8.18$$

The physical motion of the solitary wave as described by equation 8.16 is that of an interface between the low-symmetry phase and that of the high-symmetry phase. Its profile along the propagation direction is shown in Fig.8.7. Its propagation front is bent with a curvature of the form (d-1)/R for d>1.

The growth and dissolution laws depend now on the time evolution of R. It is clear from the analytical dependence of R on the dimension of the problem, that the propagation rate dR/dt in two and three dimensions approaches the propagation rate, v, in one dimension, asymptotically for large cluster sizes R.

Integrating the rate law in equation 8.13 one finds the inverse rate law:

$$t = 1/v' \{ (R-R_o) + R_c \ln(1 + (R-R_o)/(R_o-R_c))\} \qquad\qquad 8.19$$

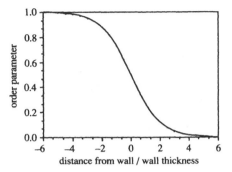

Fig.8.7 Shape of the interface between a domain with Q=0 (i.e. the high symmetry phase) and Q=Q_0 (i.e. the low symmetry form). The unity of Q is Q_0, distances are measured in units of the wall thickness.

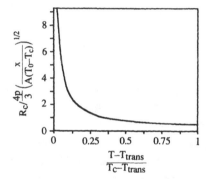

Fig.8.8 The critical radius in dimensionless units versus temperature in the interval from T_c to T_{trans} ($T_{trans} > T_c$).

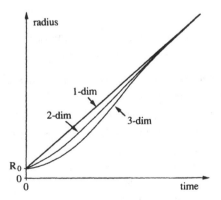

Fig.8.9 The growth law of a ferroelastic domain in a paraelastic matrix for $T < T_{trans}$ sketched for the one, two and three dimensional case.

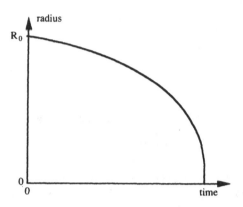

Fig.8.10 Dissolution rate law after Chan (1977).

where:

$$R_c = 1/2 \ g \ (d-1) \ \tau /v' \qquad\qquad\qquad 8.20$$

R_c is the critical radius above which there is only growth and below which there is only dissolution. The radius R_0 is determined by the starting conditions. The temperature evolution of the critical radius, R_c, is plotted in Fig.8.8, showing a critical increase at temperatures close to the transition temperature T_{trans}. The final rate laws for d=1,2,3 are depicted in Fig.8.9 as determined numerically by Chan (1977), and the rate of dissolution is shown in Fig. 8.10.

8.2 Early stages of twinning in a cubic-tetragonal phase transition and some generalisations

Transformation twins in ferroelastic and co-elastic materials are created by the phase transition from the paraelastic phase to the co-elastic phase. In the early stages of their existence the twin boundaries are highly mobile and show a finite width which leads to their description as solitary waves (see last chapter). In order to illustrate their physical behaviour in some detail, we now discuss the geometrically simple cubic to tetragonal transition as it occurs in leucite (Palmer et al. 1988, 1989), $Ba_2NaNb_5O_{15}$ (Aizu 1976) and in martensites such as Nb_3Sn, V_3Si and In-Tl alloys. We have chosen this simple geometrical model because it illustrates the essentials of the solitonic behaviour: most of the general conclusions also hold for other geometries.

Barsch and Krumhansl (1984) have treated the three-dimensional problem in their pioneering work, Jacobs (1985) achieved a full treatment in two dimensions and Falk (1983) in one dimension. Here we illustrate some of the basic physical features of the precursor structures, which eventually lead to twinning in the low-symmetry phase, and refer the reader to research articles for a more complete discussion (e.g Pouget 1988, and references given therein).

Let us start from a Gibbs free energy within the framework of the Landau-Ginzburg theory which contains three strain components:

$$e_{ik} = \begin{pmatrix} e_{11} & 0 & 0 \\ 0 & e_{22} & 0 \\ 0 & 0 & e_{33} \end{pmatrix} \qquad 8.21$$

It is convenient for further calculations to form the following symmetry adapted strain components as combinations of components of the strain tensor:

$$e_{vol} = (e_{11} + e_{22} + e_{33}) / \sqrt{3}$$

$$e_{orth} = (e_{11} - e_{22}) / \sqrt{2}$$

$$e_{tetr} = (e_{11} + e_{22} - 2e_{33}) / \sqrt{6} \qquad 8.22$$

The strain e_{vol} is a dilational strain which leads to a volume anomaly, e_{orth} is a deviatoric strain which describes the squeezing of the structure in the x-direction with a compensating expansion along the y-axis. The resulting stable structure has orthorhombic symmetry. The volume of the crystal is not changed by e_{orth}. Finally, e_{tetr} is a tetragonal lattice distortion without change of the volume of the crystal. The tensor components are defined as components of the Lagrangian strain tensor:

$$e_{ik} = 1/2 (u_{ik} + u_{ki} + u_{ji}u_{jk}) \qquad 8.23$$

and

$$u_{ik} = du_i / dx_k \qquad 8.24$$

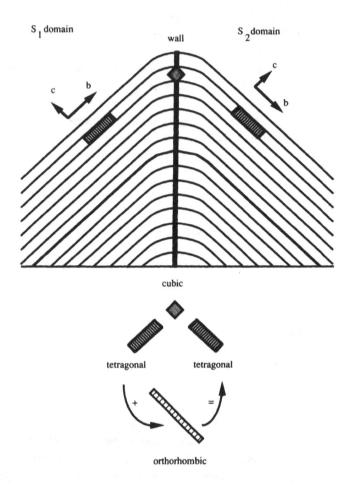

Fig.8.11 The kink interface between two tetragonal domains. The length of the a-axis is here identical to that of the c-axis. Structural elements of the paraelastic cubic phase are present in the domain wall (a=b=c). The structural elements in both domains are highlighted. The switching from an element in S_1 into S_2 can be formally achieved by adding to S_1 the spontaneous strain of the orthorhombic structure which is energetically degenerate with the tetragonal structure for $T=T_c$.

are the derivatives of the displacement vectors u with respect to the Cartesian coordinates x,y,z coinciding with the conventional cubic axes of the high-symmetry phase. Note that the strains e_{orth} and e_{tetr} transform as two basis functions of the active representation E_g. It is relevant for the understanding of the following arguments to notice that the degeneracy of the two strains e_{orth} and e_{tetr} has in fact a double meaning. For one, they are degenerate because they have, for symmetry reasons, the same energy when they occur in the cubic phase. This means that fluctuations which locally create a tetragonally distorted structure or an orthorhombically distorted structure occur with the same probability. The second meaning of this degeneracy is that the ferroelastic switching between two tetragonal domains occurs via a formal interaction with the orthorhombic strain in the low-symmetry phase. Let us illustrate this with the domain pattern depicted in Fig.8.11. The strain of the domain S_1 is:

$$e\left(S_{tetra\ 1}\right) = \begin{pmatrix} e_{11} & 0 & 0 \\ 0 & e_{11} & 0 \\ 0 & 0 & -2e_{11} \end{pmatrix} \qquad\qquad 8.25$$

with the basis function $e_{11} + e_{22} - 2e_{33}$ The strain for the second domain is, for example:

$$e\left(S_{tetra\ 2}\right) = \begin{pmatrix} e_{11} & 0 & 0 \\ 0 & -2e_{11} & 0 \\ 0 & 0 & e_{11} \end{pmatrix} \qquad\qquad 8.26$$

with the basis function $e_{11} - 2e_{22} + e_{33}$. In order to switch from S_1 to S_2 we add the strain:

$$3e\left(S_{orth\ 1}\right) = 3\begin{pmatrix} 0 & 0 & 0 \\ 0 & -e_{11} & 0 \\ 0 & 0 & e_{11} \end{pmatrix} \qquad\qquad 8.27$$

with the basis function $e_{22} - e_{33}$. We see that the strain $e(S_{orth\ 1})$ and its basis function are indeed related to the orthorhombic distortion of the lattice which had the same energy in the high-symmetry phase but no longer occurs as a static structure in the low-symmetry phase. Its role is to relate the different tetragonal states and it has the same symmetry as the switching process. Following the arguments in Chapter 5, we find that the twin boundaries in the tetragonal phase are along the {110} planes and are hence W-walls.

The total elastic energy can be written within the framework of Landau-Ginzburg theory (in order to simplify the treatment, we take e_i as the order parameter, see Chapter 5):

$$G = L(e_i) + G_{Ginzburg}$$

with the Landau energy:

$$L = \frac{1}{2} A (T\text{-}T_c)\left(e^2_{orh} + e^2_{tetr}\right) + \frac{1}{3} B\, e_{tetr}\left(e^2_{tetr} - 3\, e^2_{orth}\right) + \frac{1}{4} C\left(e^2_{orth} + e^2_{tetr}\right)^2 t \quad 8.28$$

and the Ginzburg term:

$$G_{Ginzburg} = \frac{1}{2} g \left\{ \left(\nabla_1 e_{orth}\right)^2 + \left(\nabla_2 e_{orth}\right)^2 + \frac{1}{3}\left[\left(\nabla_1 e_{tetr}\right)^2 + \left(\nabla_2 e_{tetr}\right)^2\right] \right.$$

$$+ \frac{2}{\sqrt{3}}\left[\nabla_1 e_{orth}\nabla_1 e_{tetr} - \nabla_2 e_{orth}\nabla_2 e_{tetr}\right]\Big\}$$

$$+ H\left[\left(\nabla_1 e_{tetr}\right)^2 + \left(\nabla_2 e_{tetr}\right)^2 - \sqrt{3}\left(\nabla_1 e_{orth}\nabla_1 e_{tetr} - \nabla_2 e_{orth}\nabla_2 e_{tetr}\right)\right] \quad 8.29$$

If the numerical value of B is non-zero, Landau theory leads to a first order phase transition in classical theory. The dynamical excitation of the system which eventually leads to twin formation again follows from the general Lagrange-type equation of motion (e.g. Barsch and Krummhansl 1984):

$$\rho_0 \ddot{u}_i = \sum_j \nabla_j \frac{\partial G}{\partial u_{ij}} - \sum_{jk} \nabla_j \nabla_k \frac{\partial G}{\partial u_{ijk}} \qquad\qquad 8.30$$

where

$$u_{ij} = \nabla_j u_i \text{ and } u_{ijk} = \nabla_k \nabla_j u_i$$

Within the spirit of the solutions obtained in Chapter 8.1 it is reasonable to assume that the solution will have the form of a travelling wave:

$$u_i = \alpha\, x_i + f\left(n\, x \pm v\, t\right) \quad i = \text{vol, orth} \qquad\qquad 8.31$$

$$u_{tetr} = -2\,\alpha\, x_{tetr} \qquad\qquad 8.32$$

where the constant α describes the tetragonal strain and the unknown function, f, represents the wave profile with the travelling direction, n. As the twin boundary is along {110}, say (110), we find $n = [1\bar{1}0]$. For T>>T$_c$ with α=0 and small amplitudes f and slopes f' the solution of 8.30 is a sinosoidal wave with the character of an acoustic shear wave propagating in $[1\bar{1}0]$ direction with a polarisation vector along [110]. The dispersion relation in the linearized form of 8.30 is then given by the Landau-Ginzburg parameters characterising the entropy term, A, and the gradient energy term, g:

$$\rho_o \, \omega^2 = A\,|T\text{-}T_c|\,k^2 + \frac{1}{2}\,g\,k^4 \qquad\qquad 8.33$$

The essential feature here is the additional curvature of the dispersion relationship which gives an upwards bend of the ω versus k curves for a positive gradient term g and a downwards curvature for a negative value of g. Stable solitary waves require a positive gradient energy and hence an upward curvature of the equivalent phonon branch for small wavevectors in the high-symmetry phase. This important physical relationship between the phonon dispersion at $T \gg T_C$ and the stability of solitary waves at $T \ll T_C$ was first pointed out by Barsch and Krummhansl (1984). Following their rather ingenious treatment we now illustrate the solitary waves as related to the twin-structure for $T \ll T_C$. These authors have shown that a simple solution can be obtained for the specific parameter set ($e_{vol} = 0$, $A = -2B^2/C < 0$) with a profile of the solitary wave of:

$$f(s) = f_o \pm \left(2\frac{g}{C}\right)^{\frac{1}{2}} \ln\{\cosh[s\text{-}s_o]\} \qquad\qquad 8.34$$

and a strain field of:

$$e_{orth}(s) = \pm \left[\frac{\rho_o\, v^2 + \frac{3}{4}A\,|\,T\text{-}T_c|}{\frac{1}{2}C}\right]^{\frac{1}{2}} \tanh[s\text{-}s_o] \qquad\qquad 8.35$$

where the space variable is that of the moving observer:

$$s = (n\,x \pm v\,t)/w$$

where

$$w = \left[\frac{g}{\rho_o\, v^2 + \frac{3}{4}A\,|\,T\text{-}T_c|}\right]^{\frac{1}{2}} \qquad\qquad 8.36$$

is the wall thickness at velocity v. The deformation pattern and the strain profile of the ($1\bar{1}0$) twin boundary in the ferroelastic, tetragonal phase is shown in Fig.8.12. The axes of this plot are such that the width of the boundary is w, the midpoint is on the middle of the moving or static wall and the crystallographic axes are [110] on the horizontal and [$1\bar{1}0$] on the vertical axis. The orientation of the domain wall is a vertical plane (110) and the

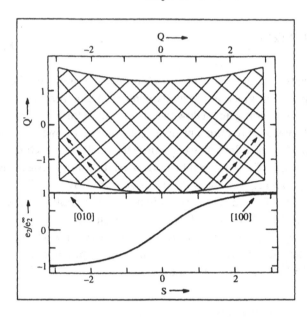

Fig.8.12 Deformation pattern (top) and strain profile (bottom) of a (1$\bar{1}$0) twin boundary in a ferroelastic material. The unrelaxed twin boundary was shown in Fig.8.5a, the structural elements were shown in Fig.8.11. The unique axis of the tetragonal phase is shown along [010] and [100] with respect to the crystallographic axes of the cubic phase (i.e. the paraelastic phase). (after Barsch and Krummhansl 1984).

propagation direction is perpendicular to the plane. The crystal structure at the left hand side of the wall approaches, with increasing distance from the wall, that of the static tetragonal phase, with the tetragonal c-axis along the cubic [010] direction. On the right hand side of the wall we find the same continuous approach to the tetragonal structure with the tetragonal c-axis along the cubic [100] axis. The solitary wave represents a switch from one orientation to the other which, as we have seen above, is equivalent to the superposition of the 'orthorhombic' strain e_{orth}. This strain field can be normalized in such a way that it disappears in the middle of the wall. The strain profile is shown in the lower part of Fig.8.12. It clearly represents a kink solitary wave which becomes a twin wall deep in the low-symmetry phase. The structural state in the wall is essentially that of the high-symmetry phase although other strain components, strictly speaking, lead to additional deformations of the lattice as shall be shown later.

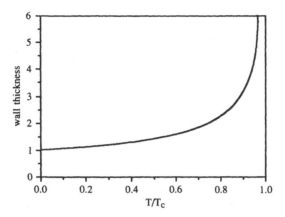

Fig.8.13 Temperature dependence of the wall thickness for a pinned domain wall. The minimum wall thickness is chosen as unity. In the case of a framework structure (e.g. feldspar) this unit is comparable with the dimension of the unit cell (say some 10Å).

The static twin wall is simply one limiting case of a kink solitary wave. In the limit of a vanishing velocity, v, of the kink the width of the boundary given by equation 8.36 becomes (for g>0):

$$w_o = \left(\frac{4}{3} \frac{g}{A|T-T_c|}\right)^{\frac{1}{2}} \qquad\qquad 8.37$$

The thickness of the twin boundary is hence, as expected from the discussion in Chapter 8.1, determined by competition between the entropy term of the Landau potential and the gradient energy in the Ginzburg energy. The wall is perfectly sharp if there is no gradient energy which could prevent the wall from collapsing. The temperature dependence of the quadratic term in the Landau potential, on the other hand, ensures that the wall thickness decreases rapidly when T approaches the absolute zero temperature. The minimum wall thickness is:

$$w_{min} = \left(\frac{4}{3} \frac{g}{AT_c}\right)^{\frac{1}{2}} \qquad\qquad 8.38$$

The wall thickness increases dramatically at temperatures close to the transition point and we find a divergence with an exponent 1/2 for a second order phase transition in the classical limit. A plot of w as a function of temperature, using some estimates of the gradient energies and the experimental values of A in the Na-feldspar structure, is shown in Fig. 8.13. It seems reasonable to assume that similar results apply for other framework minerals.

The interfacial energy of the static domain wall also follows from the energy of the solitary wave in the limit of v=0. Barsch and Krummhansl (1984) found:

$$E_o = \left(\frac{3}{2} A \mid T\text{-}T_c \mid \right)^{\frac{3}{2}} \frac{(2g)^{\frac{1}{2}}}{C}$$

8.39

This energy (per unit area of the wall) increases with increasing $\mid T\text{-}T_c \mid$ and disappears at T_c where the formation of the twin requires no additional strain energy. From a macroscopic viewpoint, the increasing ease with which domain walls can be formed when the temperature approaches the transition point is equivalent to the experimentally well known effect of an 'elastic softening' of the crystal structure with respect to two dimensional twin walls, very similar to the softening of the structure with respect to the bulk elastic constants (Chapter 5). The main difference however, is that the additional ingredient for the twin-softening is the gradient energy which does not occur in the bare Landau potential and the scaling of the energy which, in the mean field limit, decreases as $\mid T\text{-}T_c \mid^{3/2}$ instead of $\mid T\text{-}T_c \mid$ as in the cases of the elastic constants. Figure 8.14 shows the temperature evolution of the interfacial energy using the same values of g, A and C as in Fig.8.13.

We now briefly return to the correlation between the strain profile of the solitary wave and the structural state of the crystal in the solitonic region. It is sufficient to illustrate this correlation in a two-dimensional picture, including higher order strain components. We shall also consider the change of profile if we go from a second order phase transition to a first order one. This behaviour is best seen if we start from Landau potential of the form 3.6 including quadratic, fourth order and sixth order terms where the sign of the fourth order term changes from positive (i.e. second order) to negative (i.e. first order). The case of a tricritical behaviour with B=0 is discussed separately in the next chapter. The kink solitary wave in a second order phase transition has again the same profile as in equation 8.35. The displacement vectors are in linear approximation identical with those in equation 8.34. Higher order displacement vectors follow from the series expansion of u in powers of the strain field. Jacobs (1985) found the first three terms in this series for a second order phase transition to be:

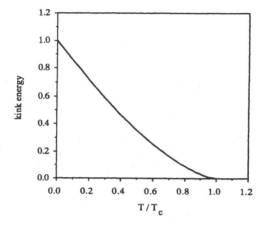

Fig.8.14 Temperature dependence of the kink energy in a second order phase transition. The energy is normalised with respect to the kink energy at zero temperature.

$$u(X,Y) = \left[1,\overline{1}\right] e\; w \ln\left[\cosh\tfrac{X}{w}\right]$$

$$-\left[1,1\right]\tfrac{1}{\sqrt{2}}\, e^2\; w\left(\tfrac{X}{w} - \tanh\tfrac{X}{w}\right)$$

$$-\left[1,1\right]\tfrac{1}{2\sqrt{2}}\, e^4\; w\left[\tfrac{X}{w} - \tanh\tfrac{X}{w} - \tfrac{1}{3}\tanh^3\tfrac{X}{w}\right] + \dots \qquad\qquad 8.40$$

where e is the equilibrium spontaneous strain.

All displacements vanish again in the middle of the wall. This point, therefore, has strictly the symmetry of the high-symmetry phase. Moving away from this point, we find that the structural deformation systematically approaches that of the low-symmetry phase but with some additional distortions parallel to the domain wall which are neither described by the tetragonal symmetry nor by the orthorhombic distortion which is degenerate at $T \gg T_c$. They are best represented by the elastic bending of the lattice planes in order to obtain a defect-free fit between the two adjacent domains. The bent lattice planes are shown in Fig.8.15 for two sets of parameters in the Landau potential. The additional low-symmetric distortions only occur in the direct vicinity of the middle of the kink and disappear rapidly in both domains with increasing distance from the kink.

We can now ask what happens to the kink if the phase transition is slightly discontinuous rather than second order. The answer is that not much is changed at temperatures far away from the coexistence regime of both phases. These are again phonon-like excitations with upward curved phonon-

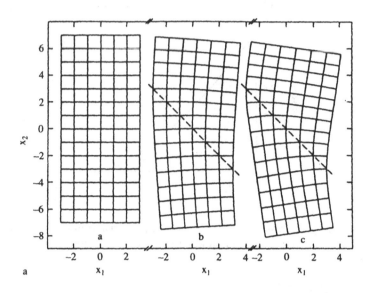

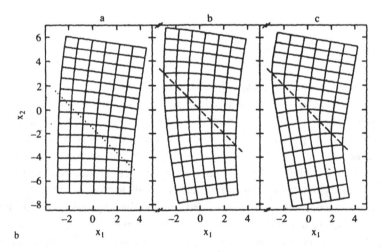

Fig.8.15a Unstrained (**a**) and strained (**b** and **c**) regions for a rectangular-rectangular soliton in the case B>0. The Landau parameters are A=-0.1 in b, A=-0.4 in c, and B=5, C=0 and g=0.2 in both. The dashed lines are the centres of the domain walls (the lines along which the strain e_2 vanishes). (after Jacobs 1985).
b Square-rectangular soliton for a: A=0.2 in a, A=0.196 in b, A=0 in c, B=-10, C=125, g=0.2 (after Jacobs 1985).

dispersion at temperatures well above the transition point. In the low temperature phase there are twin-like kinks which become real twin structures in the static limit. New phenomena develop, however, in the temperature interval where both phases are stable or metastable. New kinks evolve between regions of the high-symmetry phase and one of the possible domains of the low-symmetry phase. Their profile is related to an exponential decay of the strain with:

$$e = e_o \sqrt{1 + \exp\left(-2\frac{x}{w'}\right)} \qquad\qquad 8.41$$

and a displacement field:

$$u \propto [1,\bar{1}]\frac{1}{w'} \ln\left[\left(1 + \exp\left(2\frac{x}{w'}\right)\right)^{\frac{1}{2}} + \exp\left(\frac{x}{w'}\right)\right] + \cdots \qquad\qquad 8.42$$

(Jacobs 1985). The parameter w' is again a measure of the thickness of the domain wall. This structural variation is shown in Fig.8.16. The co-existence regime is hence characterized by two types of kinks, namely the interfaces between two domains with low symmetry (i.e. twin-like kinks) and interfaces between domains with high and low symmetry (i.e. walls of nucleation clusters). As the latter kinks show only half the possible strain variation between the two domains at large distances compared with the twin-kinks, their interfacial energy is much reduced and it becomes energetically more favourable to split one twin-kink into two nucleation-kinks (Fig.8.17). This splitting of solitons was first noted by Lajzerowicz (1981) in the case of ferroelectric materials. We would expect that this effect is even more pronounced in co-elastic materials because of the long range character of the strain fields. A twin structure in the low-symmetry phase would then broaden when the temperature is raised and the strain contrast between the two domains is simultaneously reduced. When the lower temperature limit, T_c, of the co-existence regime is reached, we expect to find a flattening out of the strain profile in the domain wall so that the high temperature phase now appears as an integral part of the wall in its middle (see Fig.8.16 and 8.17).

The slab of high temperature material is limited on either side by a smooth domain wall with either opposite orientation (a breather solitary wave) or with the same orientation (two kinks). Both situations are illustrated in Fig.8.18.

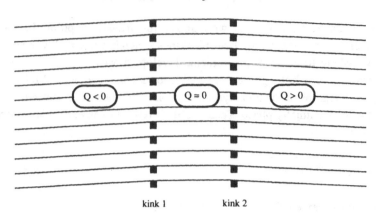

kink 1   kink 2

Fig.8.16a) Double kink configuration in a first-order phase transition. The structural
state (ferroelastic) with $Q=-Q_0$ transforms into the twin-related state $Q=+Q_0$ via an
intermediate state ($Q=0$) which is identical to the paraelastic state.

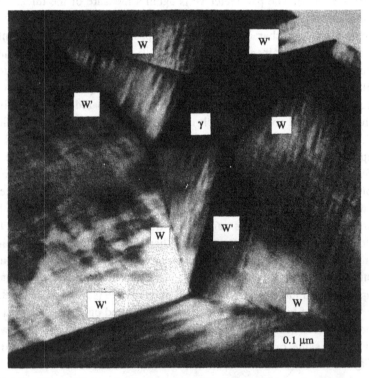

Fig.8.16b) W and W' domains in ferroelastic $Pb_3(VO_4)_2$. The centre of the
intersecting domains forms a triangular domain in which the paraelastic phase (γ) is
found representing a metastable structural state stabilised by the surrounding domain
pattern.

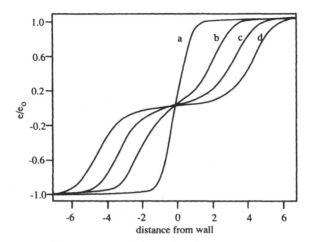

Fig.8.17a) Strain profiles near a domain wall in a first order phase transition. The strain is normalised with respect to the equilibrium strain in the ferroelastic phase. The parameters are $T = T_c$ (a), $\dfrac{T - T_c}{T_{trans} - T_c} = 0.98$ (b), 0.998 (c), 0.9998 (d) . Note the plateau at e=o for $T > T_c$ which increases with T approaching the transition temperature (after Jacobs 1985).

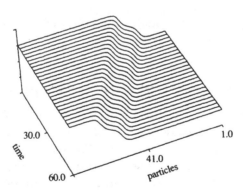

Fig.8.17b) Strain profile as in 8.17a (case c) for a moving domain wall (after Pouget 1988).

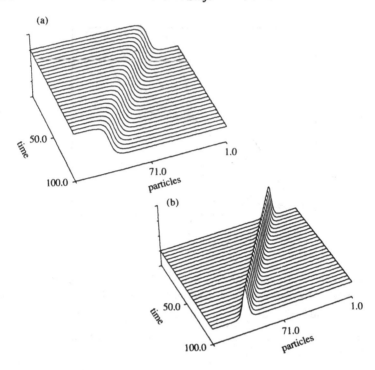

Fig.8.18 Profiles of moving kink solitons **a)** (case a in Fig.8-17a) and breather **b)**. The breather represents a ferroelastic state in a paraelastic matrix or a paraelastic state in a ferroelastic matrix (after Pouget 1988).

8.3 Domain walls in tricritical, co-elastic phase transitions

Tricritical behaviour is particularly common in co-elastic, ferroelastic and also ferroelectric phase transitions, e.g. SbSI (Samara 1978), Ag_3AsS_3 (Baisa et al. 1979), $PbZr_x Ti_{1-x} O_3$ (Clarke and Glazer 1976). Within the context of the discussion of domain structures, it is sufficient to understand a tricritical phase transition as the borderline between a second order and first order transition. We shall also give some margin to this definition in so far as phase transitions 'close to' tricritical shall also be considered. A more precise definition of tricriticality and its vicinity is given in Chapter 9.1. Typical examples of tricritical behaviour in ferroelastics are shown by crystals in the series $Pb_3(Pb_x V_{1-x} O_4)_2$: typical non-ferroelectric co-elastic materials with tricritical behaviour are calcite, anorthite, and with somewhat larger deviations $NaNO_3$. Tricritical behaviour in ferroelectrics was found in materials such as SbSI, Proustite (which is also ferroelastic), KH_2PO_4 and

$PbZr_x Ti_{1-x} O_3$ (Samara 1978, Peercy 1975, Benguigui and Beaucamps 1980, Baisa et al. 1979, Schmidt et al. 1978, Bastie et al. 1978, Clarke and Glazer 1976). All these materials have in common that their physical behaviour follows the predictions of Landau theory rather well (as discussed in Chapter 5 and 10). The relevant Landau potential for a tricitical phase transition is:

$$L(Q) = \frac{1}{2} A(T-T_c) Q^2 + \frac{1}{6} C Q^6 \qquad\qquad 8.43$$

and the Ginzburg term is identical with $1/2 \, g \, (\nabla Q)^2$, as before. The order parameter couples bilinearly with the strain and we find the same form of the Landau-Ginzburg energy in units of e as in units of Q. If we replace the potential in equation 8.28 by 8.43, and proceed in the same way as described in Chapter 8.2, we arrive at the dynamical equation (Gordon, 1983):

$$g \frac{\partial^2 e}{\partial x^2} = A(T-T_c) e + Ce^5 \qquad\qquad 8.44$$

This equation can be solved under the domain wall condition that de/dx vanishes for large distances from the domain wall. The solution is then:

$$e_{tricritical} = (A(T-T_c)/C)^{\frac{1}{4}} / \left(C \left[\frac{3}{2} coth \ 2(\frac{x}{w}) - \frac{1}{2} \right]^{\frac{1}{2}} \right) \qquad\qquad 8.45$$

where w is again the wall thickness:

$$w = \frac{1}{2} \left(\frac{g}{A \, |T-T_c|} \right)^{\frac{1}{2}} \qquad\qquad 8.46$$

We can compare this profile with that of the second order phase transition in equation 8.17 which is, in the same units:

$$e_{second \ order} = (A(T-T_c)/B)^{\frac{1}{2}} \ tanh \left(\frac{x}{w} \right) \qquad\qquad 8.47$$

Wall profiles are plotted on the same scale in Fig.8.19 and Fig.8.20.

The solutions for walls between the paraphase and the ferroelastic phase can now be calculated using the same formalism but different boundary conditions. The boundary conditions in this case are $Q=0$ for $x \to \infty$ and $Q=Q_0$ for $x \to -\infty$. The kink walls are then described by the profiles:

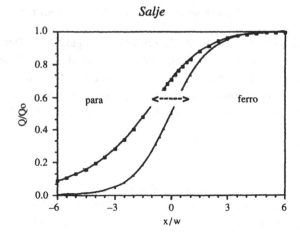

Fig.8.19 Profiles of walls between a paraelastic and a ferroelastic domain. The w for a tricritical phase transition (squares) is smoother than for a second order phas transition (line).

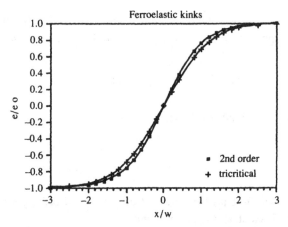

Fig.8.20 Strain profiles for walls between two ferroelastic domains in a second order and tricritical phase transition. The profiles are practically identical.

$$Q = \frac{Q_o}{1 + \exp\left(\frac{x}{w}\right)}$$

8.48

with:

$$w = \sqrt{\frac{2g}{A\,|T-T_c|}}$$

8.49

in the case of a second order phase transition and:

$$Q = \frac{Q_o}{\sqrt{1 + \exp\left(\frac{x}{w^*}\right)}}$$

8.50

with:

$$w^* = \sqrt{\frac{3}{4} \frac{g}{A|T-T_c|}} = \sqrt{\frac{3}{2}} \, w$$

8.51

in the case of a tricritical phase transition. Both walls are plotted in Fig.8.20 showing the steeper profile for the second order transition.

The interfacial energy of the domain wall is obtained from the energy deviation from the homogeneous crystal:

$$E(\text{wall}) = \int (G(x) - L(x)) \, dx$$

8.52

where $L(x)$ is the Landau energy of the homogeneous crystal. The wall energy is hence essentially the difference between the Landau-Ginzburg energy and the Landau energy:

$$E(\text{wall}) = \int \left[\frac{1}{2} A (T-T_c) \left(e(x)^2 - e^2 \right) + \frac{1}{6} C \left(e(x)^6 - e^6 \right) + \frac{1}{2} g \left(\frac{\partial e}{\partial x} \right)^2 \right] dx$$

8.53

The expectation value of e is given by the energy minimum of the Landau potential:

$$e = (A(T-T_c)/C)^{\frac{1}{4}}$$

8.54

The wall energy obtained by the integration of equation 8.53 is given by:

$$E(\text{wall}) = A (T-T_c) \left(\frac{g}{C} \right)^{\frac{1}{2}} \left[\frac{1}{12} + \sqrt{3} \ln \left| (1 + \sqrt{3}) / (1-\sqrt{3}) \right| \right]$$

$$\approx 2.4 A(T-T_c) \sqrt{\frac{g}{C}}$$

8.55

This energy increases linearly with temperature when the temperature is lowered below T_C. This can be contrasted with the behaviour in a second order phase transition where we find that the wall energy increases as

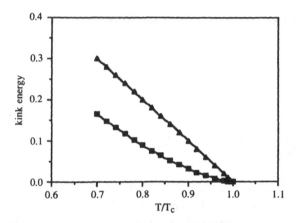

Fig.8.21 Wall energies of second order (squares) and tricritical (triangles) phase transitions. The total wall energies are normalised to unity at zero temperature. The second phase transition always has a lower wall energy than a tricritical phase transition.

$|T-T_c|^{3/2}$ (eq. 8.10). The temperature evolution of the wall energies of a tricritical and a second order phase transition are shown in Fig.8.21. Here the assumption is made that at absolute zero temperature, both energies are identical because they are only determined by the local elasticity of the crystal but not by entropy. We see that, with increasing temperature, the wall energy in the tricritical case always exceeds that of the second order material. This is important at temperatures near T_c where the wall energy vanishes asymptotically for the second order phase transition but not for the tricritical one (which is with respect to the domain formation rather similar to a slightly first order phase transition). The probability of creating domain walls at temperatures slightly below T_c is thus much larger in the case of a second order phase transition than in the tricritical case, which is in accordance with the general observation that strain-fluctuations are less relevant in a tricritical material than in a crystal with a second order phase transition.

Experimental evidence for this dramatic increase of domain walls (or the decrease of their spacings) when the transition point is approached is shown in Fig.8.22 for the ferroelastic superconductor $Co:YBa_2Cu_3O_7$, similar observations were reported by Wenyuan et al. (1985) in $InLa_{1-x}Nd_xP_5O_{14}$.

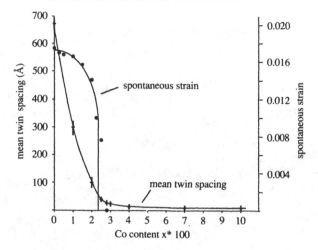

Fig.8.22 Ferroelastic phase transition in superconducting $YBa_2Cu_3O_7$: Co. The phase transition occurs at a critical Co-concentration of $c_c=2.4\%$. At $c>c_c$, the spontaneous strain disappears (tetragonal phase). The mean twin distance in the tetragonal phase reduces dramatically with increasing Co-content. The equivalent microstructures are shown in Fig.7.5.

8.4 Acoustic instabilities near wall intersections

We have described twin walls for temperatures near the transition point as kink solitary waves. In the high temperature form, we predicted an upward bend dispersion curve of the acoustic phonon branch. This phonon branch then softens when the phase transition point is approached. As the slope of the phonon branch for small wavevectors is directly correlated with the elastic constants, we can see that the softening of the phonon branches is directly correlated with the softening of the elastic constants as discussed in Chapter 6. We have seen in Chapter 7 that the twin walls W and W' in the ferroelastic phase are orientated (almost) perpendicular to each other so that we may expect that twin structures or kink-type modulations with these two preferential orientations are the 'natural' habit of a ferroelastic crystal held for a long time at temperatures just slightly below the transition point. If the kinks are mobile and intersect, we find that the preferential microstructure is the tweed pattern as shown in Fig.8.23. Experimental observations show indeed that tweed microstructures are a common effect in all co-elastic materials which twin in the co-elastic phase.

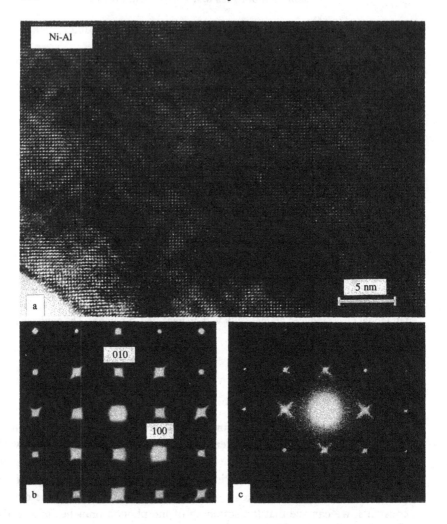

Fig.8.23 Tweed structure in Ni-Al alloy (seen along [001] a). The different areas of the structure are twisted against each other leading to the typical 'star' shaped diffraction pattern in b). The optical diffraction pattern of a) is shown in c) showing the close similarity with b). (Courtesy of Van Tendeloo and Van Landuyt).

We are now interested in the correlation between the softening of the elastic constants and the correlated phonon branches. It will be shown that this correlation leads to a simple geometrical configuration in that the critical modes have a transverse acoustic character, travelling in the twin walls with their amplitudes parallel to the junctions between the walls.

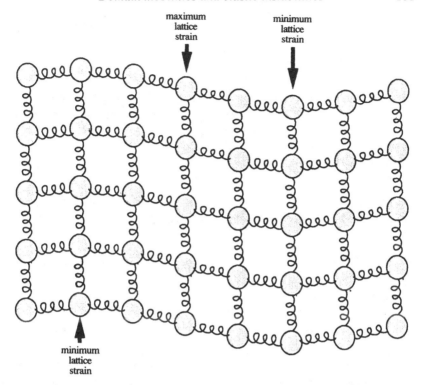

Fig.8.24 Acoustic wave in a 'ball and spring' model. The maximum deformations (i.e. the maxima of the wave) show the same structural elements as the ideal static crystal, namely square units. These areas carry, thus, the minimum lattice strain. The areas between two maxima are sheared and represent the most distorted structural elements (i.e. maximum lattice strain). The amplitude pattern of the wave and the lattice strain are thus phase shifted by half a wavelength.

Let us consider an acoustic vibration with wavenumber q and frequency $\omega(q)$. Its local displacement pattern is determined by the amplitude $u(x_j)$ at the lattice point x_j. The components of u are:

$$u_i(x_j) = u_{oi} \exp(-i(q_j x_j - \omega t))$$ 8.56

and the associated periodic strain distribution is:

$$e_{ij}(x_i) = \frac{\partial u_i(x_j)}{\partial x_j} = -i\, q_j\, u_i(x_j)$$ 8.57

A typical acoustic wave in a two-dimensional lattice is shown in Fig.8.24.

Symmetrisation of the strain tensor yields the acoustic strain tensor:

$$\eta_{ij} = -\frac{1}{2}(e_{ij} + e_{ji}) = \frac{1}{2}\,i\,(q_j\,u_i + q_i\,u_j) \qquad\qquad 8.58$$

The kinetic energy of the acoustic phonon is $1/2\,\rho\,\omega^2 u^2$, the strain energy is $-1/2\,\Sigma\,C_{ijkl}\eta_{ij}\eta_{kl}$. The total energy E_{tot} is the sum of the kinetic energy and the strain energy and we can write for $E_{tot}=0$:

$$\rho\,\omega^2\,u^2 = C_{ijkl}\,\eta_{ij}\,\eta_{kl} \qquad\qquad 8.59$$

with the elastic constants C_{ijkl} in tensor notation and the density ρ. The velocity of the acoustic vibration is defined as:

$$v = \frac{\omega}{q} \qquad\qquad 8.60$$

which leads with eq. 8.59 to:

$$\rho\,v^2 = C_{ijkl}\,\eta_{ij}\,\eta_{kl}/(qu)^2 \qquad\qquad 8.61$$

If the phase transition is co-elastic with the formation of a spontaneous strain, the elastic constants soften as described in Chapter 5 and the matrix C_{ijkl} shows eigenvalues which, in bilinear coupling, tend to zero at the transition point. The frequency of the coupled phonon follows in its temperature evolution, that of the elastic constants and we expect these modes to slow down at the transition temperature. This 'acoustic softening' has been observed in many ferroelastic crystals and is illustrated in Fig.8.25 for $Pb_3(PO_4)_2$.

The properties of the soft acoustic mode can now be calculated from equations 8.61 and 8.58. Bearing in mind the symmetry of the elastic tensor $C_{ijkl}=C_{jikl}=C_{ijlk}=C_{jilk}$ we find:

$$\rho\,\omega^2\,u_i\,u_k = \Sigma\,C_{ijkl}\,q_j\,q_l\,u_i\,u_k \qquad\qquad 8.62$$

We can now simplify this equation and introduce the 'dynamical matrix':

$$M_{ik}(q) = \underset{jl}{\Sigma}\,C_{ijkl}\,q_i\,q_k \qquad\qquad 8.63$$

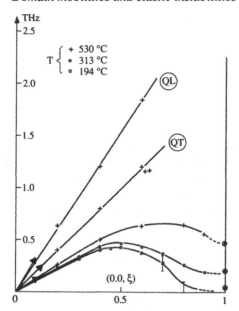

Fig.8.25 Acoustic softening in $Pb_3(PO_4)_2$. The phase transition occurs at 180°C at the L point of the Brillouin zone (x=1). On lowering the temperature from T=530°C, the acoustic phonon branch does not only soften at the critical point but also reduces the slope at the origin. This slope represents the wave velocity v and is directly related to the elastic constants $v \approx \sqrt{\dfrac{c_{66}}{\rho}}$ (Joffrin et al. 1979).

so that the equation 8.62 becomes:

$$\rho \, \omega^2 \, u_i \, u_k = M_{ik} \, u_i \, u_k \qquad\qquad 8.64$$

The condition that the acoustic mode becomes soft is then identical to the condition that the dynamical matrix has vanishing eigen-values. As M_{ik} is a symmetric and real matrix, all eigen functions are orthogonal. The critical modes have, thus, orthogonal amplitudes and propagation directions.

The relationship between the acoustic strain and the spontaneous strain have already been discussed by David (1983). It is easy to verify that the acoustic strain is proportional to the difference between the elastic strain components of two domain orientations, m and n:

$$\eta_{ij} = \frac{1}{2\sqrt{2}} [(e_s)_{ij} (m) - (e_s)_{ij} (n)] \qquad\qquad 8.65$$

It is now interesting to reformulate the strain-free condition of the domain wall in Chapter 7, using the dynamical acoustic strain:

$$[(e_s)_{kj}(m) - (e_s)_{kj}(n)] x_i x_j \propto \eta_{ij} x_i x_j = 0 \qquad 8.66$$

The latter identity shows that the domain wall directions and the displacements of the acoustic modes are identical. As the domain walls W and W' are (almost) perpendicular to each other, we also find that their directions agree with the propagation direction of the acoustic soft modes.

We can now take up the example of Na-feldspar in Chapter 7 to illustrate the geometrical relationship for the soft modes and the domain walls. The elastic stiffness tensor of a monoclinic feldspar is in Voigt notation:

$$C_{ik} = \begin{pmatrix} c_{11} & c_{12} & c_{13} & 0 & c_{15} & 0 \\ & c_{22} & c_{23} & 0 & c_{25} & 0 \\ & & c_{33} & 0 & c_{35} & 0 \\ & & & c_{44} & 0 & c_{46} \\ & & & & c_{55} & 0 \\ & & & & & c_{66} \end{pmatrix} \qquad 8.67a$$

Following the arguments in Chapter 5, Salje (1985) has shown that the combination of the elastic constants which becomes critical in Na-feldspar is:

$$c_{44}\, c_{66} - c_{46}^2 \rightarrow 0 \qquad 8.67b$$

with the related critical components of the strain tensor e_4 and e_6. The dynamical matrix can then be written for a mode with the wavevector:

$$q = [0 \; q_y \; 0]$$

as

$$M_{ij} = q_y^2 \, C_{iyjy} \qquad 8.68$$

The non-zero components C_{iyjy} are in Voigt notation c_{66} (i=x,j=x), c_{64} (i=x,j=z), c_{44} (i=z,j=z) and c_{22} (i=y,j=y). The dynamical matrix is then:

$$M_{\alpha\beta} = q_y^2 \begin{pmatrix} c_{66} & 0 & c_{46} \\ 0 & c_{22} & 0 \\ c_{46} & 0 & c_{44} \end{pmatrix} \qquad 8.69$$

The two solutions for the velocities in equation 8.64 are the longitudinal wave with velocity:

$$v_{l_1} = (c_{22}/\rho)^{\frac{1}{2}}$$

8.70

and the two transverse waves with velocities:

$$\rho v_{t_{1,2}}^2 = \frac{c_{44} + c_{66}}{2} \mp \left[\left(\frac{c_{44} + c_{66}}{2}\right)^2 - c_{44}\,c_{66} + c_{46}^2\right]^{\frac{1}{2}}$$

8.71

Only one of these velocities, namely v_{t1}, becomes zero if the instability condition (equation 8.67b) is fulfilled. The amplitude of this mode is:

$$\begin{pmatrix} c_{66} & c_{46} \\ c_{46} & c_{44} \end{pmatrix} \begin{pmatrix} x \\ z \end{pmatrix} = \rho\, v_{t_1}^2 \begin{pmatrix} x \\ z \end{pmatrix}$$

8.72

with

$$v_{t_1} = \left(-\frac{c_{46}}{c_{66}}, 0, 1\right)$$

8.73

at the point of the phase transition. This mode has therefore all the properties required from the soft acoustic mode, i.e.:

a The wavevelocity vanishes at the point of the phase transition (in this quasiharmonic approximation);
b The mode propagates in the Pericline wall because its wavevector points along the crystallographic b-axis (see Chapter 7) and;
c The amplitude is perpendicular to the wall. Its orientation is therefore within an Albite twin wall if one intersects with the Pericline wall.

It is now easy to show that a second critical acoustic vibration occurs with a wavevector in the Albite wall and an amplitude perpendicular to the Albite wall (see Salje et al. 1985a for details). The main geometrical features of these modes are illustrated in Fig. 8.26.

Finally, it is interesting to discuss these acoustic instabilities within the context of wall intersections as discussed in Chapter 7. It follows from equation 8.63 that the two propagation directions are perpendicular to the junction between the twin walls and also to each other. The amplitudes are always parallel to the second domain wall. The phase between both modes shall now be fixed at a common origin in the junction (Fig.8.26).

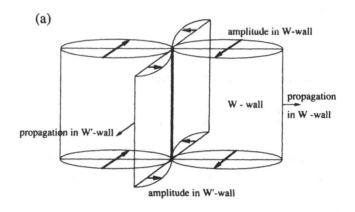

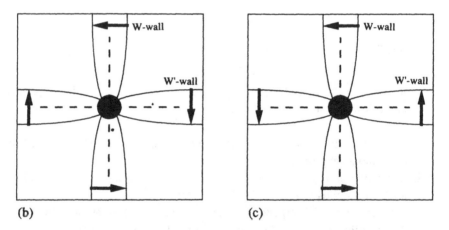

Fig.8.26 Dynamical behaviour of soft acoustic vibrations near T_C. Their orientation is related to the twin walls W and W' at $T<T_C$. The phonon amplitudes are perpendicular to the wall planes (a). Interlocked phonons follow the wall deformation around the junction $\left(\frac{\pi}{2} \pm \omega,\ \text{see Fig.7.8}\right)$ (b) or rotate around the junction (c). The periodic superposition of these movements leads to a dynamical tweed pattern similar to Fig.8.27.

The quasiharmonic vibration then has the same characteristics as a shear motion (a) or a rotation (b) of the lattice around the junction. Although one might now intuitively assume that this lattice distortion is energetically unfavourable it appears that there are crystals where this shear or rotation motion leads to a decrease of the total energy. It is this reduction of energy

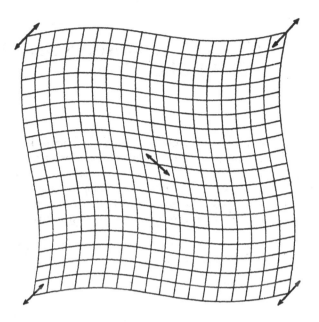

Fig.8.27 The superposition of the deformation pattern shown in Fig.8.26 illustrating the effect of two simple transverse distortion waves on an initially orthogonal lattice point array. The arrows indicate the orientations of the regions of maximum distortion. This pattern was first used by McConnell (1971) for the description of the microstructure in adularia.

and its static equivalent of additional shear distortions of a lattice which is the basic idea of the McConnell model of the modulated phase in adularia where a second lattice of repeated junctions is superimposed on to the original lattice of the monoclinic feldspar structure (Fig.8.27) (McConnell, 1965,1968,1971,1975). We shall return to the general discussion of negative coupling energies in Chapter 9.

9

SPECIFIC HEAT ANOMALIES AND THE EXCESS ENTROPY

Investigations of the temperature evolution of the excess entropy lead to a quantification of the order parameter behaviour on a macroscopic scale. The excess entropy is usually determined from measurements of the specific heat over a range of temperatures. Landau theory is now applied rather forcefully including the discussion of the role played by defects and coupling phenomena. Mineralogists may appreciate the applications to the feldspar structure.

We have seen that one of the basic assumptions of Landau theory as introduced in Chapter 3 was the direct correlation between the excess entropy ΔS and the order parameter Q:

$$\Delta S \propto Q^2 \qquad\qquad 9.1$$

If ΔS also depends on higher correlations as reflected by Q^4, Q^6 etc., we find that the higher order coefficients of the Landau potential B, C etc. also depend on temperature. This extension of the temperature dependence of $G(Q,T)$ as discussed so far, does not change any of the results obtained in the previous chapters. It makes it, however, much more difficult to analyse experimental results in a quantitative way because the number of parameters involved in the analysis increases rather dramatically. In some cases, the temperature dependence of the higher order terms has been measured directly from structural data (e.g. Na-feldspar, anorthite) and no significant temperature evolution of the higher order terms was found (Salje et al. 1985b, Redfern et al. 1988). This experimental observation might thus serve as a phenomenological justification of equation 9.1 in ferroelastic and co-elastic crystals.

The most direct way to determine the excess entropy due to the phase transition is via the measurement of the excess specific heat ΔC_p as a function of temperature:

$$\Delta S\ (T) = -\int_T^{T_c} \frac{\Delta C_p}{T}\ dT \qquad\qquad 9.2$$

The direct experimental determination of ΔC_p also has the advantage that it allows us to assess the significance, in thermodynamic terms, of the Ginzburg free energy as related to fluctuations. At temperatures close to the transition temperature of a continuous phase transition, fluctuations of the order parameter lead to extra contributions to the specific heat as:

$$\delta \Delta C_p \sim \int \frac{V}{T} \langle Q_k Q_{-k} \rangle^2 \, dk \qquad\qquad 9.3$$

where V is the volume and Q_k is the k-th component of the space dependent order parameter Q(r). The bracket $\langle\rangle$ indicates statistical averaging. The temperature dependence of C_p near T_C allows, in principle, for direct measurement of the integrated order parameter fluctuations. Moreover, remembering that the introduction of critical indices was largely motivated by the experimental observation of stronger singularities in C_p at T_C than expected from Landau theory, namely:

$$\Delta C_p \propto |T\text{-}T_c|^{-\alpha} \qquad\qquad 9.4$$

the central interest in investigating the behaviour of C_p, related to structural phase transitions in general, becomes obvious. The main experimental techniques for the determination of ΔC_p are discussed in some detail by Salje (1988b), here we analyse the significance of such data for the characterisation of a co-elastic phase transition.

9.1 Landau theory and specific heat anomalies

Starting from the most simple Landau expression (e.g. 3.6):

$$L(Q) = \frac{1}{2} A'Q^2 + \frac{1}{4} BQ^4 + \frac{1}{6} CQ^6 \qquad\qquad 9.5$$

with $A' = A(T\text{-}T_c)$ we find the excess entropy simply as:

$$\Delta S = -\frac{\partial L}{\partial T} = -\frac{1}{2} AQ^2 \qquad\qquad 9.6$$

In a second order phase transition with C = 0, the excess specific heat is a linear function of temperature with a step at T_c:

$$\Delta C_p (T) = T \frac{\partial \Delta S}{\partial T} = \frac{A^2}{2B} T, \quad T \le T_c \; ; \; \Delta C_p (T) = 0, \, T \ge T_c \qquad\qquad 9.7$$

In a discontinuous phase transition with $C \neq 0$ the excess entropy and the excess specific heat are:

$$\Delta S = \frac{A}{4C}\left[B - \sqrt{B^2 - 4AC\,(T\text{-}T_c)}\right]$$

9.8

$$\Delta C_p = \frac{A^2 T}{2\sqrt{B^2 - 4AC\,(T\text{-}T_c)}}$$

9.9

If the first order phase transition is close to a tricritical point with small values of B, the difference between the temperature T_0 at which $B^2 = 4A'C$ and T_c, i.e.:

$$T_c - T_0 = -\frac{B^2}{4\,A\,C}$$

9.10

tends to zero as the square of B. The singularity of C_p then becomes in the lowest order:

$$\Delta C_p = T\left(\frac{A^3}{16C}\right)^{\frac{1}{2}}\frac{1}{\sqrt{T_0 - T}}$$

9.11

Thus the specific heat increases near T_c of a tricritical phase transition or a 'Gauss model' (Ma 1976) as:

$$\Delta C_p \propto (T\text{-}T_c)^{-\alpha}, \quad \alpha = 0.5$$

9.12

The experimental data on $NaNO_3$ show values of α close to 1/2. The phase transition $P\bar{1}$-$I\bar{1}$ in highly ordered anorthite (Val Pasmeda) in Fig.9.1 is also tricritical with $\alpha = 0.5$ (Wruck 1986). Helwig et al. (1977) observed near-tricritical points in $AgNa(NO_2)_2$, and other $\alpha = 0.5$ singularities were reported by Shang and Salomon (1980).

A convenient representation of C_p in slightly first order phase transitions follows from equation 9.9 and 9.10 as:

$$\left(\frac{T}{\Delta C_p}\right)^2 = \frac{4\left(B^2 - 4AC(T_0\text{-}T_c)\right)}{A^4} + \frac{16\,C}{A^3}(T_0 - T)$$

9.13

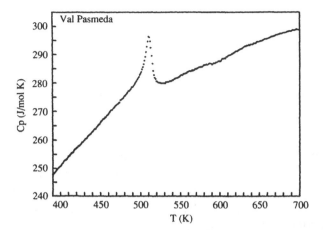

Fig.9.1 Specific heat anomaly in anorthite from the Val Pasmeda locality. This crystal has a high degree of Al,Si ordering. The anomaly is related to the $I\bar{1}$-$P\bar{1}$ transition and follows the behaviour of a tricritical phase transition (courtesy B. Wruck).

which is a linear function of temperature. The first term gives the discontinuity of the specific heat at the transition temperature T_o.

From equation 9.13 the relative magnitude of the parameters A, B and C in the Landau potential can be determined. In a second order phase transition, the last term in equation 9.13 vanishes and the first is reduced to $4B^2/A^4$ as in equation 9.7. The closeness of a phase transition to the tricritical point or to a second order phase transition can therefore be estimated from the $(T/\Delta C_p)^2$ versus T plots (e.g. 9.13). Experimental data are discussed by Aleksandrov and Flerov (1979), and Aleksandrov et al. (1985). The development of excess specific heat, without consideration of the additional contributions of fluctuations, is therefore expected to be a smooth function of B when the order of the phase transition is changed. A single step is predicted from Landau theory for a continuous phase transition, a $|T-T_o|^{-1/2}$ - singularity at the tricritical point and a discontinuity for first order phase transitions. Although the Landau potential with short range interactions is an unrealistic approximation of the Gibbs free energy near a transition line of the second order, it is noteworthy that the conditions for this approach to be valid are more easily satisfied as the tricritical point is approached and only logarithmic corrections are expected for small values of B.

The most universal way to determine the free parameters in the Landau potential follows from equation 9.8 which can be rewritten as:

$$\Delta S = \frac{2}{3} \frac{L}{T_{trans}} \left\{ 1 + \left[1 - \frac{3}{4} \frac{T - T_c}{T_{trans} - T_c} \right]^{\frac{1}{2}} \right\}$$ 9.14

or

$$\frac{4}{3} \left[-\left(\frac{\Delta S}{2L/3T_{trans}} - 1 \right)^2 + 1 \right] = \frac{T - T_c}{T_{trans} - T_c}$$ 9.15

T_{trans} is the temperature at which both phases are in thermodynamic equilibrium, L is the latent heat. If the expression on the left hand side of equation 9.15 is plotted against temperature, the intersection with the temperature scale defines T_c. The slope of the curve is $(T_{trans} - T_c)^{-1}$ which allows the determination of T_{trans}.

An alternative Landau polynomial with second-, third-, and fourth-order terms gives the following expression for the excess entropy:

$$\Delta S = \frac{3}{4} \left(\frac{L}{T_{trans}} \right) \left\{ 1 + \left[1 - \frac{8}{9} \left(\frac{T - T_c}{T_{trans} - T_c} \right) \right]^{\frac{1}{2}} \right\}$$ 9.16

which can be fitted to the experimental data points for the determination of T_c and T_{trans}. Differences in the excess entropies expressed by equations 9.15 and 9.16 therefore allow the distinction between first order phase transitions with negative fourth order terms and those involving symmetry breaking third order terms. A typical example is the $P2_13 - P2_12_12_1$ symmetry reduction in $K_2Cd_2(SO_4)_3$ where third order terms are allowed from symmetry (Dvorak 1972). Heat capacity measurements by Devarajan and Salje (1984), however, demonstrated that their numerical values are nonetheless small and the first order character of the phase transition results from a large, negative fourth order term.

9.2 Influence of Fluctuations of the Order Parameter on the Specific Heat Anomalies

Close to T_c significant fluctuations with long wavelengths occur and Fourier components with small k-values become important. We have seen in Chapter 8 that these fluctuations are highly restricted in reciprocal space and that significant correlations occur in the fluctuations. Let us first, however, consider a random fluctuation which we shall subsequently restrict. The temperature interval of the fluctuation regime depends sensitively on the 'interaction length' L_c, which characterises the range of interaction causing the structural phase transition. The Ginzburg criterion relates the interaction length, the temperature interval ΔT of the fluctuation regime and the excess specific heat ΔC_p which should be expected if the Landau theory were valid as follows:

$$\Delta T = \frac{k_B^2 \, T_c}{32\pi^2 (\rho_m \Delta C_p)^2 \, L_c^6}$$

9.17

where ρ_m is the molar density.

It seems likely that the interaction length ranges from one interatomic distance up to several lattice constants, the corresponding ΔT would then vary between some tens of degrees down to vanishingly small intervals. Equation 9.17 can, on the other hand, be used for the estimation of the interaction length if ΔC_p and the temperature interval in which additional fluctuation induced excess specific heats appear are known from calorimetry. In all co-elastic materials we expect $L_c > 50\text{Å}$ which leads with $\rho_m \Delta C_p = a^6 k_B$, where $a \sim 10\text{Å}$ is the unit cell dimension, to an extremely small Ginzburg interval of $\Delta T / T_c \sim 10^{-8}$, which is experimentally not accessible.

For a vanishing conjugate field H and $T > T_c$, the averaged order parameter is zero and the effective Hamiltonian becomes, in the lowest order:

$$H_{eff} = V \sum_{k<k_o} \left(\chi^{-1} + \frac{1}{2} g k^2 \right) |Q_k^2|$$

9.18

The quadratic dependence of H_{eff} on Q_k implies a Gaussian distribution in the integral 8.1. For such integrals the mean value of the product of several Q_ks is identical to the sum of products of pairs of mean values of the correlation functions of the fluctuations in k-space. As Levanyuk (1963) has shown, the integral in equation 8.1 using the effective Hamiltonian in 9.18 gives:

$$G = -k_B T_c V \sum_{k<k_o} \ln \frac{\pi T}{\left(\chi^{-1} + \frac{1}{2} g k^2 \right)^2}$$

$$= k_B T_c V \int_o^{k_o} \ln \frac{\chi^{-1} + \frac{1}{2} g k^2}{\pi T} \frac{2\pi k^2}{(2\pi)^3} dk$$

9.19

Under the assumption that this correction holds equally for $G_{Ginzburg}$ in 8.3, the resulting fluctuation correction for the specific heat becomes:

$$\partial C_p = C_p - C_P^O - C_P^Q = \frac{VA^2T_c^2}{16\pi^2} \int_o^\infty \frac{k^2 dk}{\left(\chi^{-1} + \frac{1}{2} gk^2\right)^2}$$

$$= \frac{V\left[\frac{1}{2} AT_c\right]^{\frac{3}{2}}}{16\pi \left[\frac{1}{2} g\right]^{\frac{3}{2}}} \frac{\sqrt{T_c}}{\sqrt{|T-T_c|}} \propto \frac{1}{\sqrt{|T-T_c|}} \qquad\qquad 9.20$$

This power law is identical for tricritical behaviour (equation 9.12) and, as will be shown later, is also identical to certain defect-induced excess specific heats.

We have seen that the isotropic Hamiltonian (9.18) can be considerably changed in ferroelastic and co-elastic crystals because of the effect of elastic fields. As a result, not all fluctuation waves increase near T_C in sufficiently anisotropic crystals, and this leads to a strong decrease of the phase volume in equation (9.19).

A typical example is the behaviour of C_p near T_C if bilinear coupling between Q and the local elastic strain e is allowed as in Chapter 5. Starting from the effective Hamiltonian of the form (5.2):

$$H_{eff} = \frac{1}{2} A(T-T_c) Q^2 + \frac{1}{4} BQ^4 + \frac{1}{2} g(\nabla Q)^2 + \sigma Qe + H_{el} \qquad\qquad 9.21$$

Levanyuk and Sobyanin (1970) found the resulting pair correlation function:

$$\langle Q_k Q_{-k}\rangle = \frac{k_B}{V\tilde{\alpha}(k)} \qquad\qquad 9.22$$

where $\tilde{\alpha}(k)$ is the renormalised coefficient of the quadratic term in the effective Hamiltonian (Levanyuk and Sobyanin 1970):

$$\tilde{\alpha}(k) = \chi^{-1} + \frac{1}{2} gk^2 - \sigma^2 \left(\frac{1}{4\mu} - \frac{1}{4\mu}\left(\cos^2\theta + \frac{\lambda+\mu}{\lambda+2\mu} \sin^4\theta \sin^2 2\phi\right)\right) \qquad\qquad 9.23$$

in the case of an elastically isotropic material. The susceptibility χ is defined in equation 5.5. The first two terms are identical to the isotropic, uncoupled Hamiltonian, the last term stems from the elastic energy H_{el} and acts as damping. The specific heat correction due to these fluctuations becomes:

$$\delta C_p = C_p - C_P^O - C_P^Q \propto \int \frac{dk}{\tilde{\alpha}^2(k)} \qquad\qquad 9.24$$

which converges for $T \rightarrow T_c$. Only the first derivative with respect to $(T-T_c)$ becomes singular, with:

$$\frac{\partial \delta C_p}{\partial T} \propto |T-T_c|^{-\frac{1}{2}}$$

9.25

The singularity in C_p is accompanied by cusps in C_v according to:

$$C_v \propto C_{vo} - \chi'/C_p \,, T \rightarrow T_c$$

9.26

with constants C_{vo} and χ'. If C_p increases as the power law $|T-T_c|^{-\alpha}$, only $\partial C_v/\partial T$ becomes singular with:

$$\partial C_v/\partial T \propto |T-T_c|^{-\alpha-1}$$

9.27

9.3 The Influence of Lattice Imperfections

Most minerals and materials used for industrial applications contain a great number of lattice imperfections which may change the physical behaviour of the crystal in a major way. It is well known that impurities, as one prominent example of lattice imperfections, play a crucial role in the performance of superconductors, semiconductors and ionic conductors. In the case of ferroelastic and co-elastic materials one finds a less dramatic dependence on the density of most lattice imperfections because of the nature of the long-ranging strain fields. Most impurities will build up local strain fields which interact with the order parameter in the same way as described in Chapter 5. The major difference is, however, that the local strain is inhomogeneous and hence leads to space-dependent order parameters in the sense of Landau-Ginzburg behaviour rather than pure Landau theory. Starting from the idea that the order parameter variation close to the lattice imperfection follows the Ornstein-Zernike function we may write for the defect induced order parameter:

$$Q_{defect} = \sum_{lm} Q_{c,lm} \frac{1}{r} e^{-\frac{r}{r_c}} \psi_{lm}(\theta,\phi), \quad r > d$$

9.28

with the correlation length r_c of the order parameter Q. The amplitude Q_c is a measure of the magnitude of the lattice distortion and d is the diameter of the hard core of the defect. The orientational dependence is reflected by the angular functions $\psi_{lm}(\theta,\phi)$ which represent spherical harmonics. In general, the expansion in equation 9.28 is truncated after the first isotropic function where the lattice distortion is an S-wave. This is, however, a physically rather unsatisfactory assumption because we have seen that the elastic instability of the crystal due to the structural phase transition is highly anisotropic with two elastically soft directions in the lattice. The most probable local lattice distortion will follow this anisotropic pattern, with the angular dependence of Q_{defect} being related to dipolar and, most importantly, quadrupolar distortions.

The effective Hamiltonian is then of the type:

$$H_{eff} = \int dV \left\{ \frac{1}{2} A(T\text{-}T_c) Q^2 + \frac{1}{4} BQ^4 + \frac{1}{6} C Q^6 + \frac{1}{2} g \left(\nabla Q \right)^2 - H_d(r) Q(r) \right\} \quad 9.29$$

where the conjugate field is created by the defects:

$$H_d(r) = \Sigma \; \xi_{pq} \, Q_c^p \, Q^{q-1} \quad\quad\quad 9.30$$

leading to a coupling energy:

$$H_d(r) Q(r) = \Sigma \; \xi_{pq} \, Q^q(r) \, Q_c^p(r) \quad\quad\quad 9.31$$

For small defect concentrations we can ignore the defect-defect interactions (i.e. terms with $q \geq 2$). The strongest interaction might be due to bilinear coupling (i.e. $p = q = 1$) which has already been discussed in detail in Chapter 5. The result of defects are in this case the renormalisation of T_c and its spread over a wider temperature range:

$$T_c^* = T_c - \int \left\{ \xi \frac{1}{r} e^{\frac{r}{r_c}} \psi \right\} dV \quad\quad\quad 9.32$$

We see here already that if the lattice relaxes the critical correlation length r_c diverges for $T \rightarrow T_c$ in a second order phase transition and the effective area of the crystal which is affected by the imperfection becomes identical with the lattice. At the other extreme, the influence of lattice imperfections for $T \ll T_c$ is rather small and limited to the hard core ($r \leq d$). In Fig.9.2 the relative influence of impurities on the order parameter for negative and

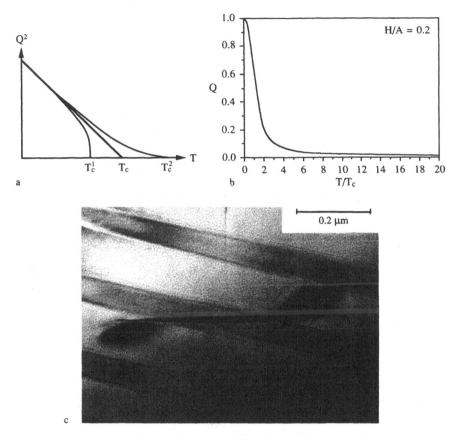

Fig. 9.2a) Temperature dependence of the order parameter on the presence of defects in bilinear coupling. The linear dependence with the transition temperature represents the defect free crystal in a second order phase transition. For non-symmetry breaking defects, the transition temperature is lowered to T_c^1 and the critical exponent near T_c^1 is smaller than the classical value. If the defects are symmetry breaking, the transition temperature is increased to T_c^2. In the case of homogeneous fields H_d, T_c^2 is infinity. For weaker, inhomogeneous fields, T_c^2 is a finite temperature. The order parameter curve of the symmetry breaking defects is often referred to as having a 'defect tail' at $T > T_c$.

b) Temperature dependence of the order parameter in 2-4 potential and a strong, homogeneous defect field H=0.2A. Experimentally this means that 'parasitic' superstructure reflections remain visible over large temperature intervals above T_c.

c) Example of the interaction of the order parameter and extended defects. The dark lenticular area represents a needle-shaped domain with W-walls (albite twin in peristerite). The extended bands are exsolution lamellae. The twin walls are broken and off-set at the walls of the exsolution lamellae, the strong strain fields close to the tips of the needles appear as dark areas in the image (Courtesy of W.L. Brown, Nancy).

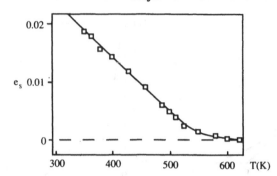

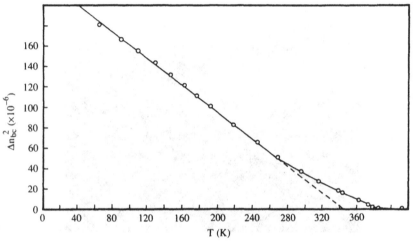

Fig.9.3 a) Defect tail in $Pb_3(AsO_4)_2$. The spontaneous strain is proportional to the square of the order parameter. The defect free transition has a second order phase transition. This crystal shows $T_c^2 \approx T_c + 50K$.
b) Temperature dependence of Q^4 ($\propto \Delta n_{bc}^2$, the morphic birefringence) in Ba doped $Pb_3(PO_4)_2$ $(Pb_{0.95} Ba_{0.05})_3 (PO_4)$. The phase transition is nearly tricritical and shows a defect related tail at $T > 260K$. (Courtesy of U. Bismayer, Habilitationsschrift, 1989).

positive coupling are sketched and in Fig.9.3 some examples of Ba-doped $Pb_3(PO_4)_2$ and impure $Pb_3(AsO_4)_2$ are shown. Similar renormalisation of T_c was found by Gurskas et al. (1981) in the case of Li-doped rare earth molybdates. These authors also observed an additional phase transition at lower temperatures than the ferroelastic transition point, which occurs only in doped crystals but not in pure materials. It is not clear, however, whether this transition represents an additional potential lattice instability of the pure phase (which does not lead to a true phase transition because of insufficient thermodynamic stability) or whether non-uniform states including interplanar 'weak links' exist in these layer structures.

There are very few theoretical predictions concerning the analytical form of the defect contributions to the excess specific heat after the first pioneering work of Halperin and Varma (1976). Levanyuk et al. (1979) showed that in the case of spatially fixed, randomly distributed defects, the C_p-corrections can be calculated within the scope of Landau theory, yielding for $l = 0$, $m = 0$ in equation 9.28:

$$\delta\Delta C_p^{defect} = \pi N\, r_c^3 \left(\frac{Q_0}{Q^*}\right)^2 \Delta C_p^o \tag{9.33}$$

for $T > T_c$, and:

$$\delta\Delta C_p^{defect} = [1+4\pi N\{r_c^3\left(\frac{Q_0}{Q^*}\right)^2\left[1-2\frac{Q_\infty}{Q_0}\frac{r_c}{d} - 3\frac{Q_\infty^2}{Q_0^2}\right]$$

$$+ 4d^3\left[\ln\frac{r_c}{d} + const\right]\}]\Delta C_p^o \tag{9.34}$$

for $T < T_c$.

ΔC_p^o is the jump of C_p according to Landau theory, r_c is again the correlation radius, Q_0 is the defect induced value of Q near the defect position, Q^* is the maximum value of Q including fluctuations where $Q^* = \sqrt{2g/Bd^2}$, and d is the diameter of the defect and N is the defect concentration. Q_∞ is the solution of the undisturbed effective Hamiltonian including the field correction. The strength of the defect can be expressed by the ratio Q_0/Q^*. For strong defects with $Q_0/Q^* \approx 1$ and $Q_0 \gg Q$, the expressions 9.33 and 9.34 simplify to:

$$\delta\Delta C_p^{defect} \propto r_c^3 \tag{9.35}$$

If the critical radius depends on temperature as $r_c \sim |T-T_c|^{1/2}$, the excess specific heat shows a singularity at T_c with:

$$\delta\Delta C_p^{defect} \sim |T-T_c|^{3/2} \tag{9.36}$$

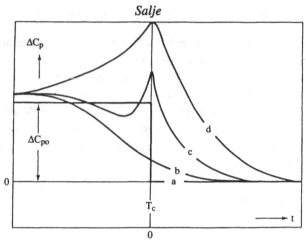

Fig.9.4 The anticipated influence of defects on the excess specific heat in the region of applicability of the Landau theory of a second order phase transition: **a** in the absence of defects, and with frozen-in random defects. The shape of ΔC_p depends on the interactions strength; **b** weak interactions cause a smearing of the phase transition; **c** intermediate interactions with an additional maximum at T_c and **d** strong interactions leading to a λ-shaped excess specific heat (after Levanyuk et al. 1979).

This temperature dependence superimposes the influences of fluctuations and may grossly modify the critical exponent of the 'ideal' crystal.

For weak defects, equations 9.33 and 9.34 describe smeared phase transitions as sketched in Fig.9.2. The actual critical behaviour in a crystal with lattice imperfections depends on the type of imperfections. For non-mobile, frozen-in defects with a statistical distribution, Levanyuk et al. (1979) found (for S-type defects):

$$\delta \Delta C_p^{\text{defect}} \propto |\ T-T_c|^{\nu-2} \qquad\qquad Q_0 \neq 0 \qquad\qquad\qquad 9.37$$

$$\delta \Delta C_p^{\text{defect}} \propto |T-T_c|^{2\beta-2} \qquad\qquad Q_0 = 0 \qquad\qquad\qquad 9.38$$

For unpolarised point defects, the excess specific heat (P-type defects):

$$\delta \Delta C_p^{\text{defect}} \propto |T-T_c|^{-\frac{1}{2}} \qquad\qquad\qquad\qquad\qquad\qquad 9.39$$

shows the same temperature dependence as for fluctuations in the limit of equation 9.20 and the tricritical behaviour in equation 9.11 (Fig.9.4).

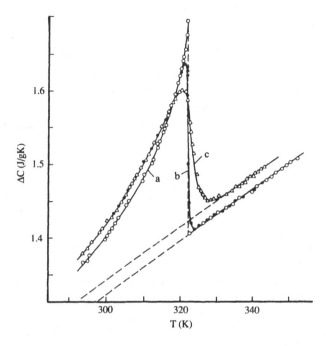

Fig.9.5 Temperature dependence of the specific heat in pure triglycine sulfate (TGS) **a** crystals doped with 3% Cr^{3+} in solution **b** and with 10% a-alanine in solution **c**. The effect of defects is to blur the transition and to change the baseline for the curve **c**. The experimental observations follow the theoretical predictions following 9.29 (after Strukov et al. 1980).

In a systematic experimental approach, Strukov et al. (1980) showed that the influence of defects on the excess specific heat of TGS crystals tends to follow the predicted pattern. The form of the C_p anomaly near $T_c = 322.18K$ depends heavily on the crystal history (Fig.9.5). The 'purest' crystals are those grown above T_c but annealed below T_c. Equilibrium distribution of the defects can be assumed in this case and the anomaly in C_p is stepwise and rather sharp. On the other hand, crystals annealed above T_c show, during their first cooling across T_c, a λ-shaped singularity at slightly reduced maximum temperatures, i.e. in the asymmetric phase if the crystal were in thermodynamical equilibrium. The onset of the increase of C_p in the high symmetry phase is identical for both experiments.

Additional defects were induced by γ-irradiation. For small doses, there is an increase in the specific heat and again a shift of T_c towards lower temperatures. If these curves were interpreted as being due to critical fluctuations, reasonable curve fittings using the expression 9.4 would have been possible with only slightly 'exotic' critical exponents, α. This example demonstrates clearly the danger of interpreting peaks in the C_p curves at T_c invariably as fluctuation-induced critical phenomena without further consideration of the possible influence of lattice imperfections. Misinterpretations of this kind are even more likely to occur because the power law singularities of C_p that are due to defect structures are expected only for low defect concentrations, whereas high defect concentrations are more likely to smear the phase transition.

A typical example for such behaviour is KH_2PO_4 where the C_p anomaly has been known since 1969 (Reese) and interpreted in terms of a power law singularity with $\alpha = 0.5$ for $T < T_c$. At temperatures above T_c, a logarithmic singularity was analysed by Reese (1969). However, Courtens (see Scott et al. 1982) showed that this anomaly can be removed entirely by proper sample preparation and annealing and is therefore due to lattice imperfections. Another prominent example of the influence of defects on specific heat measurements is the behaviour of $BaMnF_3$ where Scott et al. (1982) found a mean field discontinuity of 0.4 cal/mol K at $T_c = 255.3K$. A power law divergency of 6.4 ± 0.5 cal/mol is superimposed on top of this discontinuity and is entirely related to defects. The power law singularity satisfies the predictions of the defect dominated dynamical theory of Levanyuk et al. (1980) and was further confirmed by comparison of ultrasonic attenuation and soft mode line-width divergencies with its predictions. The influence of dislocations has been investigated by Dubrovsky and Krivoglaz (1979).

The sensitivity of C_p to defect structures can also be a starting point for systematic investigations of defect structures themselves. The pre-eminence of the heat capacity as a sensitive detector and indicator of dopant level impurities has been demonstrated for the Verwey transition in Fe_3O_4 by Bartel and Westrum (1976). A sharp transition peak appears for pure magnetite with a small high temperature shoulder. Both singularities depend sensitively on dopants, such as Zn, Cd and Mn, and show a large splitting of the two critical temperatures for $Zn_{0.005}Fe_{2.995}O_4$. The attribution of these singularities to structural and electronic phase transitions is not yet fully clarified and strong coupling effects are likely to occur.

We finally comment on the interaction between defects. We have seen that the critical radius r_c of the deformation cloud around each lattice imperfection grows when the transition temperature is approached. In the classical limit of a second order phase transition we can write:

$$r_c = r_{c_0} \left(\frac{T_c}{|T-T_c|} \right)^{\frac{1}{2}}$$

9.40

In a ferroelastic matrix, the order of magnitude of the minimum diameter of the cloud is probably at least 20Å (T = 0K) which leads to $r_c \sim$ 40Å (T = 1/2 T_c) and $r_c \sim$ 200Å (T = 0.9 T_c). Comparing these values with a typical unit cell diameter of 5Å, we find that r_c reaches the average distance between the defects at concentrations of one defect per 64 unit cells (T = 0K), one defect per 512 unit cells (T = 1/2 T_c) and one defect per 64 000 unit cells (T = 0.9 T_c). This clearly demonstrates that in co-elastic materials even small defect concentrations lead to interactions between the defects, at least at temperatures close to T_c. If the overlaps between the deformation clouds are large enough to cover the entire lattice, we can envisage the defects to be an integral part of the structure and their role can be described by a secondary order parameter (Chapter 10). In the intermediate range we can estimate the variation of the order parameter Q due to the defects as an average over the volume for isotropic, bilinear coupling:

$$= n \frac{4\pi}{V} |\overline{Q} - Q| \frac{1}{8\pi g} r_c^2 \left(1 + \frac{d}{r_c}\right)$$

9.41

where g is the coefficient of the Ginzburg energy, and $\overline{Q}$ is the value of the order parameter on the hard core of the defect. The contribution of this hard core to the total defect volume has been ignored for $r_c \gg d$. We can now estimate the order of magnitude of the change of the order paramter due to defects by using the approximation $g \sim a^2$ where a is the lattice constant and all energies are measured in units of $k_B T_c$. The equation 9.41 can then be simplified to:

$$\langle \Delta Q \rangle \sim \frac{n}{2V} |\overline{Q} - Q| \left(\frac{r_c}{a}\right)^2 \left(1 + \frac{d}{r_c}\right)$$

9.42

Let us now estimate the change of Q by using some typical values for the coefficients. The interaction length at some 0.8 T_c may be $r_c \sim$ 50Å and a $\sim$ 10Å for a framework structure. The hard core may be d $\sim$ 5Å (e.g. polyhedra diameter). We then find:

$$\frac{\langle \Delta Q \rangle}{|\overline{Q} - Q|} \sim 13.7 \frac{n}{V}$$

9.43

so that one defect per 100 unit cells which stabilises the paraelastic phase (i.e. $\overline{Q} = 0$) lead to a relative decrease of Q by 13.7%. Continuous defect fields will build up when the local defect fields (9.41) overlap, i.e. $n_c \sim 3V/4\pi r_c^3$ at $T \ll T_C$. Let us assume $r_C \sim 30\text{Å}$ at $T = 0.6\, T_C$ so that $n_c \sim 0.88\%$. At $n < n_c$, the defect fields are essentially local and lead to a smearing of the transition whereas for $n > n_c$ the extrapolated critical temperature will shift as a function of the defect concentration (see Fig.9.6).

We expect, therefore, that defects with a density of less than roughly 1% act as individual centres which smear the phase transition but do not significantly change the extrapolated transition temperature. For higher defect concentrations, first bilinear and then linear-quadratic coupling between the defect concentration and the order parameter will lead to a rather well-defined transition temperature which is now an explicit function of the defect concentration. The resulting phase diagram is depicted in Fig.9.6 where Tc does not change for $n \ll n_c$ but depends on n for $n \gg n_c$.

We have concentrated, so far, on the question of how the defects influence the structural phase transition in a co-elastic or ferroelastic material. We will now discuss the inverse correlation, namely the effect of the phase transition on the defects. As the chemical potential μ of the defects depends explicitly on the order parameter Q:

$$G_{\text{defect}} = \mu_{\text{eff}} \cdot N \propto N\left(\mu_0 + \int \Sigma \zeta_{pq} Q^p(r) Q_c^q(r)\, dr^3\right) \qquad 9.44$$

the chemical potential no longer has the value of the bare potential μ_0 which would occur without the influence of the phase transition. The direct effect of structural distortions is thus a change in the energy of the lattice imperfection and, therefore, its solubility. The partition function of the defects becomes in simple bilinear coupling:

$$Z = N \exp\left[\frac{\mu}{k_BT}\right] = N \exp\frac{\mu_0}{k_BT} \cdot \exp\left\{\frac{1}{k_BT} \xi^* Q\right\} \qquad 9.45$$

where $\xi^* = \zeta Q_c$ is the effective coupling coefficient. For large defect concentrations, we expect the defects to exsolve once a critical concentration is surpassed. Let us consider a typical situation as depicted in the phase diagram in Fig.9.6. Here we have ignored all fluctuation effects and assume that the structural phase transition is continuous. If a crystal with defect con-

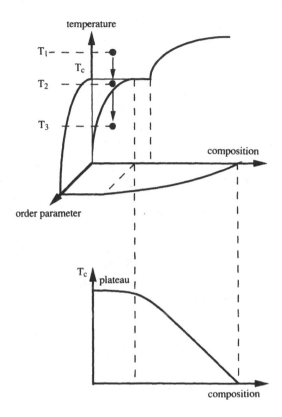

Fig.9.6 Order parameter-temperature-composition diagram (top) and composition dependence of the transition temperature (bottom). A crystal is in the paraelastic phase at T_1, all defects are dissolved without mutual interaction. Cooling the crystal to $T_2 < T_c$ leads to a structural phase transition and a reduction of the defect solubility. The exsolution curve is, however, not surpassed and the defects are weakly interacting. At T_3 the defects exsolve, often on domain boundaries (see text). The phase diagram T_c *versus* compositions shows a plateau for small concentrations with non-interacting, then weakly interacting defects. Only at higher defect concentrations, the critical temperature changes approximately linearly in the case of the most common linear-quadratic coupling $Q^2 n$.

centration n_0 is held at temperature T_1, we find that all defects are dissolved and the material is in the paraelastic phase. Cooling the sample to temperature T_2 leads to twinning in the co-elastic phase with a much reduced solubility of defects. At temperature T_3, the solubility line has been crossed on cooling and defects exsolve. The intrinsic microstructure at which these defects will cluster are the twin boundaries. This is energetically favourable because the middle of the twin boundary (in the limit of a kink solitary wave, see Chapter 8) is, indeed, identical to or similar to the structure of the paraelastic phase in which the solubility of defects is higher. When the equilibrium line $n_{critical}(T)$ is reached, defects can start to segregate on the twin boundaries. Such 'decorated' walls are thus pinned and become less mobile than equivalent walls in a non-exsolved crystal. When the crystal is reheated at temperatures above T_c for a short time, the twins will disappear instantaneously but the decoration dissolves much more slowly and will act as pinning centres for kinks if the crystal is cooled again at temperatures below T_c. This behaviour is one of the major mechanisms for the well-known memory effect of ferroelastic transformation twins, i.e. the reappearance of twin walls at exactly same position after heating a sample to $T > T_c$ and subsequent cooling into the ferroelastic phase.

A typical example is the twinning of anorthoclase with low K-content. The twin structures are those described before for Na-feldspar (Chapter 7). The Or-component of anorthoclase is, however, metastably dissolved in the alkali-feldspar and tends to exsolve at low temperatures. If a twinned crystal with triclinic symmetry is heated to about 300°C, i.e. below the inversion temperature, we find that the alkali ions are sufficiently mobile for some K to segregate to the twin boundaries. If this crystal is now heated for a short time at temperatures above T_c and quenched, it will show basically the same domain structure as before because of their fixation by K-decoration.

9.4 Order Parameter Coupling and Excess Specific Heat

We have seen in Chapter 8 that the excess specific heat and the excess entropy, being volume quantities, integrate over all temperature dependent contributions to the free energy regardless of a certain order parameter being primary, secondary, or temperature dependently coupled. Most scattering experiments and optical parameters such as birefringence and optical activity, on the other hand, detect specifically the temperature dependence of the interacting order parameter only.

Significantly different behaviour can therefore be expected if the thermal evolution of $<Q>$ determined from specific heat measurements is compared with results from these other experimental techniques.

An example for additional excess entropies due to order parameter coupling is the ferroelastic phase transition in $Pb_3(PO_4)_2$ at $T_c = 453.6K$ (Salje and Wruck 1983). The symmetry reduction R3m-C2/c is described by an effective Hamiltonian (Salje and Devarajan 1981):

$$H_{eff} = \frac{\alpha}{2}\left(Q_1^2 + Q_2^2\right) + \frac{\alpha'}{2} Q_3^2 + \left(Q_1^2 + Q_2^2\right)^2 L_2 + Q_3^2\left(Q_1^2 + Q_2^2\right)L_3 +$$

$$+ Q_3^4 L_1 + \frac{1}{\sqrt{2}} Q_1 Q_3\left(Q_3^2 - \frac{1}{3} Q_2^2\right)L_4 \qquad\qquad 9.46$$

which can be diagonalised under the assumption of well separated fixed points of the two sets of order parameters $\{Q_1, Q_2\}$ and Q_3. The order parameter Q_3 alone describes a classical second order phase transition with a Landau expression of equation 9.5. The parameters Q_1 and Q_2 follow the Hamiltonian of a Potts-oscillator with:

$$H_{potts} = \left(\frac{\alpha}{2} + L_3 M^2\right)\left(Q_1^2 + Q_2^2\right) + L_2\left(Q_1^2 + Q_2^2\right)^2 + \omega Q_1\left(Q_2^2 - \frac{1}{3} Q_1^2\right) \qquad 9.47$$

Structurally, the pseudo spin part of equation 9.47 describes the orientation of the monoclinic binary axis of the low temperature phase with respect to the trigonal unit cell of the high temperature phase (Bismayer et al. 1982). The related excess entropy follows from the leading terms in equation 9.46:

$$-\Delta S = \frac{\alpha_1}{2}\left(Q_1^2 + Q_2^2\right) + \frac{\alpha_2}{2} Q_3^2 \qquad\qquad 9.48$$

with:

$$\alpha_1 = \frac{\partial\alpha}{\partial T} \text{ and } \alpha_2 = \frac{\partial\alpha'}{\partial T} \qquad\qquad 9.49$$

The excess entropy can be compared with the optical birefringence induced by the structural phase transition (Bismayer et al. 1982; Salje and Wruck 1983):

$$\Delta n_{bc} = P_{eff}\left(\frac{1}{2}\tilde{D}^2 + 2\tilde{G}^2\right)^{\frac{1}{2}}\left(Q_1^2 + 2\sqrt{2}\,|Q_1 Q_3|\right) \qquad\qquad 9.50$$

with constant parameters $\tilde{D}$ and $\tilde{G}$.

In the displacive limit ($Q_3 = -1/\sqrt{2}\,Q_1$), the excess entropy and the morphic birefringence depend quadratically on the same order parameter and they are therefore proportional to each other. At temperatures close to T_c, this proportionality is broken, based upon differences in the temperature dependencies of the order parameters Q_1 and Q_3. In Fig.9.7 the linear relation holds for $T < 415K$, whereas larger entropy values are found for temperatures

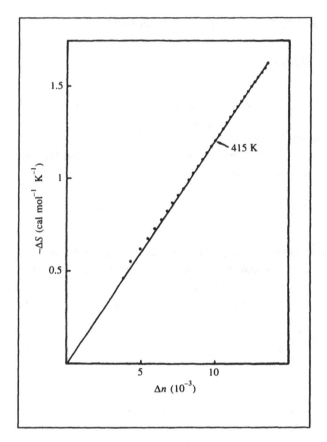

Fig.9.7 Excess entropy ΔS versus birefringence Δn_{bc} in $Pb_3(PO_4)_2$ in the temperature range 377.5-453.6K (T_c=453.6K) (after Salje and Wruck 1983).

closer to T_C. The deviation from proportionality is plotted as a function of temperature in Fig.9.8 which shows the additional divergency of $|\Delta S|$ at T_c.

This method of re-scaling the excess entropy against other macroscopic variables can also be used for investigating the order of coupling effects. One illustration is from the cubic to orthorhombic ($P2_13$ - $P2_12_12_1$) phase transition in crystals with Langbeinite structure (e.g. $K_2Cd_2(SO_4)_3$, $K_2Ca_2(SO_4)_3$). In this case multicomponent order parameter behaviour combines with reasonably strong order parameter strain coupling (Dvorak 1972; Maeda 1979; Speer and Salje 1986). The relevant strain components x_i and the morphic birefringence Δn follow from minimization of the Landau potential in a stress free crystal and are higher polynomials of the order parameter components (Devarajan and Salje 1984).

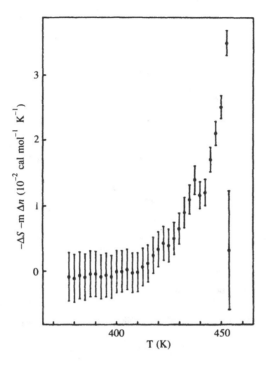

Fig.9.8 Difference between the experimentally observed excess entropy ΔS and the extrapolated value for a one-order parameter behaviour ($Pb_3(PO_4)_2$, Salje and Wruck 1983).

The general assumption, however, that the lowest order terms dominate the temperature dependence of x_i and Δn turned out to be wrong when these two quantities were plotted as functions of the correlated excess entropy (Fig.9.9 and 9.10). The higher order polynomial terms lead to strong non-linearities and there is no extrapolated common zero point of simultaneously vanishing excess entropy and lattice strain. The structural interpretation of these findings is based on the X-ray observations of Speer and Salje (1986), and Percival et al. (1989) and spectroscopic evidence (Percival and Salje 1989), where two different types of structural deformations were found to dominate the transition process. Firstly, the octahedra complexes around the divalent ions distort and expand and this is compensated by a denser packing of oxygen around the potassium positions. Secondly, the SO_4 tetrahedra tilt in order to follow the structural deformations of the larger polyhedra along the trigonal axes. Both movements contribute differently to the excess entropy,

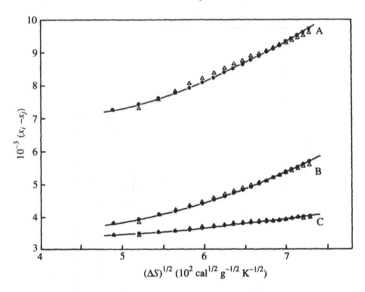

Fig.9.9 Plot of the components of the spontaneous strain versus the root of the excess entropy, which is proportional to the thermodynamic order parameter in $K_2Cd_2(SO_4)_3$ (after Devarajan and Salje 1984). A: X_2-X_3, B: X_1-X_2, C: X_1-X_3.

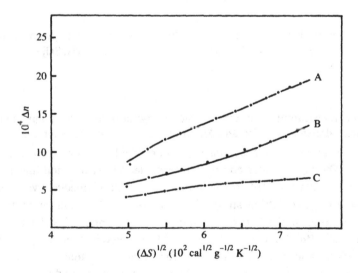

Fig.9.10 Plot of the optical birefringence in the three principal directions versus the root of the excess entropy ($K_2Cd_2(SO_4)_3$ (after Devarajan and Salje 1984). A: $\Delta n_{\alpha\gamma}$, B: $\Delta n_{\beta\gamma}$, C: $\Delta n_{\alpha\beta}$.

the ferroelastic strain and the morphic birefringence resulting in a strong non-proportionality between these parameters.

Coupling between the strain parameter and the order parameter was also reported to cause differences in the critical behaviour of ΔC_p near T_c between single crystals and crystal powders (Hatta et al. 1977). This effect is mainly due to the macroscopic shape of large single domain crystals in which the local relaxations and fluctuations are clamped by their surrounding lattice. In powder samples, the additional contribution of inhomogeneous strain fields to the individual grains may appear and smear the phase transition. Experimentally, Hatta et al. (1977) observed a peak in the ΔC_p curve at T_c for a single domain crystal of $SrTiO_3$. The critical exponent was found to be 0.14, which is close to the Ising value. On the other hand, a polydomain sample showed only a small anomaly near T_c which was attributed to the intrinsic strain free C_p value.

9.4 1.a Na-Feldspar

We now return to the discussion of the thermodynamic behaviour of feldspars as a prime example of co-elastic material with several interacting order parameters. The Gibbs free energy of the Na-feldspar can be expressed using the two order parameters Q and Q_{od} in bilinear coupling (Chapter 8):

$$G = \frac{1}{2}aQ^2 + \frac{1}{4}bQ^4 + \frac{1}{6}cQ^6 + \frac{1}{2}a_{od}\,Q_{od}^2 + \frac{1}{4}b_{od}\,Q_{od}^4$$

$$+ \frac{1}{6}c_{od}\,Q_{od}^6 + \xi Q((A + C/\chi)Q + (B + D/\chi)Q_{od})$$

$$+ \xi^{od}\,Q_{od}\,((A + C/\chi_{od})Q + (B + D/\chi_{od})Q_{od})$$

$$+ \frac{1}{2}C_{44}(AQ + BQ_{od})^2 + \frac{1}{2}C_{66}(CQ + DQ_{od})^2$$

$$+ C_{46}(AQ + BQ_{od})(CQ + DQ_{od}) + \lambda QQ \qquad\qquad 9.51$$

The thermodynamic equilibrium of Na-feldspar is described by:

$$\partial G/\partial Q = 0 \qquad\qquad 9.52$$

$$\partial G/\partial Q_{od} = 0 \qquad\qquad 9.53$$

leading to two basic equations for the two order parameters:

$$a_o(T-T_c^*)\,Q + bQ^3 + cQ^5 + dQ_{od} = 0 \qquad\qquad 9.54$$

$$a_{od,o}(T-T_c^{od*})\,Q_{od} + b_{od}Q_{od}^3 + c_{od}Q_{od}^5 + dQ = 0 \qquad\qquad 9.55$$

The entire thermodynamic behaviour of Na-feldspar depends, therefore, only on seven independent parameters, and these can all be experimentally determined. Feldspars are often found with non-equilibrium structural states because the two order parameters act on extremely different time scales. The lattice distortions, described by Q, are dynamically induced by phonon processes with typical relaxation times of faster than 10^{-10} sec. The order parameter Q can rapidly follow any change of external parameters, such as temperature or pressure. No metastability is expected for Q. The Al,Si ordering, on the other hand, is described by Q_{od}, and changes of Q_{od} require the breaking of tetrahedral (Al-O and Si-O) bonds. For each exchange of Si and Al between sites, a threshold energy is needed and, in simple cases, this energy is the activation energy for ordering.

Metastable states are described by:

$$a_o(T-T_c^*)\,Q + bQ^3 + cQ^5 + dQ_{od} = 0 \qquad\qquad 9.56$$

where Q_{od} measures the frozen-in degree of Al,Si order. The parameters a_o, b, c, d, $a_{od,o}$, b_{od}, and c_{od} have been determined experimentally by Salje et al. (1985b). The total free energy in equation 9.51 can, therefore, be expressed in terms of one order parameter alone and all related thermodynamic functions can be calculated directly: the excess specific heat of the displacive part - in a metastably fully ordered sample shows a step at T_C and the excess entropy increases at $T \ll T_C$ linearly with decreasing temperature (Fig.9.11). In thermodynamic equilibrium, both ordering mechanisms take place in the structural phase transition. The resulting excess specific heat, as in the lower part of Fig.9.11, again shows a step at T_C but is now followed by a large peak near the onset temperature of the Al,Si ordering. This peak, although similar to a λ-peak, does not indicate any symmetry breaking and is not related to a structural phase transition but is due to the coupling of the two order parameters. This example illustrates that in the case of coupled order parameters the actual phase transition is not necessarily accompanied by major changes of the excess specific heat. Such changes may well occur at much lower temperatures in the asymmetric phase without being related to a further phase transition process, however.

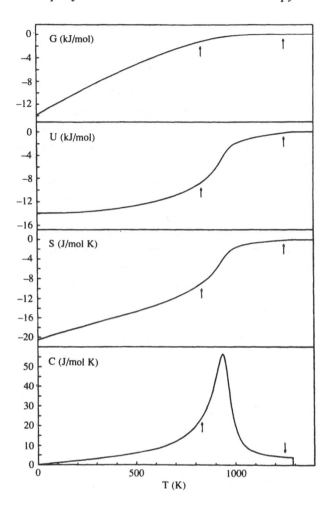

Fig.9.11 Temperature dependence of the excess free energy ΔG, excess internal energy ΔU, excess entropy ΔS and excess heat capacity ΔC for thermal equilibrium. The arrow at $T_c{}^*=1251K$ indicates the temperature of the monalbite-analbite transition. The arrow at $T_c^{od^*}=833K$ shows the temperature at which Al,Si order would start without coupling with the 'displacive' order parameter Q (after Salje et al. 1985).

9.4.1.b Ca-Feldspar

In anorthite, the coupling between the displacive order parameter and the Al,Si ordering parameter follows a different symmetry pattern. The hypothetical Al,Si ordering phase transition takes place at high temperatures above the melting point. Its symmetry change is according to our basic assumption of a monoclinic paraelastic phase C2/m → C$\bar{1}$ → I$\bar{1}$. Near 514.5K an observable phase transition between phases with the space groups I$\bar{1}$ and P$\bar{1}$ takes place. The underlying model for the displacive transition is similar to that of Adlhart et al. (1980), where the driving force of the structural phase transition is essentially identified as the condensation of a soft mode at the boundary of the first Brillouin Zone of the I$\bar{1}$ structure. In an extension of this model, clustering and ordering of alkali positions in multi-cell potentials were allowed for by Salje (1987), and this restabilises the soft mode at temperatures close to the transition temperature T_c. Fluctuations have been observed by Ghose (1989). The symmetry reduction during the phase transition is restricted to a breaking of the translational symmetry $\tau = [111]$ of the I$\bar{1}$ structure which leads to additional super-lattice reflections (h + k = even, l = odd and h + k = odd, l = even) to appear in the P$\bar{1}$ phase. The critical point of the phase transition is called the 'Z-point', with order parameter Q_Z.

For symmetry reasons, direct coupling between Q_{od} and Q_Z is not allowed. (Their respective active representations are not identical). However, the following three coupling mechanisms are consistent with the symmetry reductions of Q_{od} and Q_Z:

(i) Strain induced coupling as described in Chapter 8. The Gibbs free energy due to coupling has the form:

$$\Delta G = \Sigma \, \lambda_i \, e_i \, Q_Z^2 + \Sigma \, \zeta_i \, e_i \, Q_{od}^2 \qquad\qquad 9.57$$

(ii) Direct biquadratic coupling with:

$$\Delta G = \xi_{12} Q_Z^2 Q_{od}^2 \qquad\qquad 9.58$$

which was advocated by Anisimov et al. (1981) and discussed in Chapter 8.3. No experimental evidence for this coupling mechanism has yet been reported in the case of anorthite. The resulting thermodynamic behaviour, however, is identical with case (i).

(iii) Inhomogeneous symmetry breaking. The homogeneity of the crystal can be perturbated by fluctuations of Q_{od} and gradient coupling might occur as shown in Chapter 8.

The Gibbs free energy of a uniform, Ca-rich plagioclase includes the free energies of both order parameters, their coupling and the renormalisation of the critical temperature due to Na-Ca exchange. The expression given by Salje (1987) is:

$$G = \frac{1}{2}\alpha(T - (T_c + \gamma n_{Ab}))Q_Z^2 + \frac{1}{4}B_{eff}Q_Z^4 + \frac{1}{6}cQ_Z^6 + \Sigma\lambda_i e_i Q_Z^2 +$$

$$+ \frac{1}{2}\Sigma C_{ik}e_i e_k + \Sigma \zeta_i e_i Q_{od}^2 + \xi_{12}Q_Z^2 Q_{od}^2$$

$$+ \frac{1}{2}\alpha_{od}\left(T - T_c^{od} - n_{Ab}\frac{\partial T_c^{od}}{\partial n_{Ab}}\right)Q_{od}^2 + \frac{1}{4}b_{od}Q_{od}^4$$

$$+ \frac{1}{6}c_{od}Q_{od}^6 + \xi_{od}^* n_{Ab}Q_{od} \qquad\qquad 9.59$$

where n_{Ab} is the mole fraction of the Albite in the plagioclase, T_c is the critical temperature of the $I\bar{1}$ - $P\bar{1}$ phase transition and e_i is the i-th component of the spontaneous strain of the $P\bar{1}$ phase.

The equivalent temperature dependence of the order parameter of the Al,Si ordering ($C\bar{1}$ - $I\bar{1}$ phase transition) follows from:

$$G_{od} = \frac{1}{2}\alpha_{od}\left(T - T_c^{od} - n_{Ab}\gamma_{od}\right)Q_{od}^2 + \frac{1}{4}b_{od}^*Q_{od}^4 + \frac{1}{6}c_{od}Q_{od}^6 \ ;$$

$$\partial G/\partial Q_{od} = 0 \qquad\qquad 9.60$$

This equation expresses the fact that the phase transition takes place by changes of temperature, chemical composition or both.

The major difference between this free energy expression and the equivalent one for Na-feldspar is that the coupling between the two order parameters is bilinear for Na-feldspar and biquadratic for anorthite. The biquadratic coupling including the role of stress fields changes the critical temperature T_c of the $I\bar{1}$ - $P\bar{1}$ phase transition <u>and</u> the fourth order term of the Landau expression. This latter renormalisation implies that Al,Si disorder changes the order of the phase transition and the transition temperature.

The specific heat anomaly of the $I\bar{1}$ - $P\bar{1}$ transition in anorthite has been observed using differential scanning calorimetry (DSC) techniques by Wruck (1986). Highly ordered anorthite from the Val Pasmeda locality shows a large λ-peak with a critical exponent $\alpha = 0.5$. In samples which have either been heat-treated ($Q_{od} \ll 1$) or contain some Albite component (e.g. the anorthite from the Monte Somma locality), the fourth order term in the potential (9.60) becomes renormalised and the phase transition is second order. The λ-anomaly disappears and a smeared-out stepwise increase of the excess heat capacity was found by Wruck (1986) (see Fig.9.1).

10

COUPLING BETWEEN ORDER PARAMETERS IN FERROELASTIC AND CO-ELASTIC CRYSTALS

Ferroelastic and co-elastic crystals often undergo cascades of phase transitions with several interacting order parameters. Their interaction is the origin of 'cross-over' behaviour, and may also stabilise phases which can only exist if several order parameters couple. A descriptive introduction leads to a quantitative formulation of general cases which the reader might like to use for his or her own research work.

We have seen in Chapter 5 how the physical mechanism of ferroelastic and co-elastic phase transitions can be described by order parameters. These order parameters couple with the spontaneous strain and thereby lead to the excess lattice distortions which are an essential feature of co-elastic materials. The order parameters were treated in most examples as one-dimensional quantities, the only exception so far was the cubic-tetragonal transition discussed in Chapter 8 where the order parameter had two competing components, and we shall come back to the question of multi-dimensional order parameters in an example below. First we wish to concentrate on another aspect of interacting order parameters which is rather crucial to the understanding of the behaviour of crystals with complex structures - an effect only discovered and appreciated over the last decade or so.

It is an empirical observation that phase transitions in minerals and other materials are often driven simultaneously by more than one physical mechanism. This has led us in Chapter 6 to the definition of 'impure' ferroelastics and co-elastics as materials in which the phase transition is brought about by the interaction between more than one order parameter. Typical examples are: the cation ordering of Al and Si in feldspars; the atomic displacements involving the tilt of lattice complexes (e.g. As_2O_5 in Chapter 4, rotation of crankshafts in Na-feldspar in Chapter 9, tilt of octahedra in the perovskite structure etc.); the molecular disorder in calcite and many others (e.g. Flochen et al. 1985 on the competition between ferroelectricity and ferroelasticity, Toledano (1979) and Yoshihara et al. (1985) on similar coupling in Benzil). If a crystal structure has sufficient complexity it will

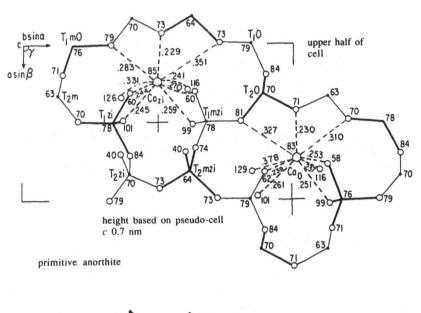

height based on pseudo-cell
c 0.7 nm

primitive anorthite

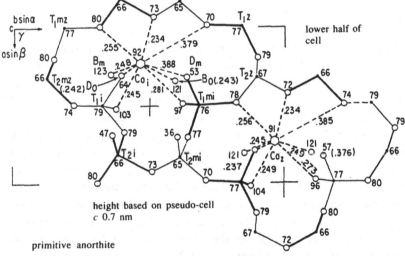

height based on pseudo-cell
c 0.7 nm

primitive anorthite

Fig.10.1a) Upper and lower halves of 1.4nm cell of primitive anorthite. Atomic coordinates from Kempster et al. (1962), and distances from Kalus (1978). Al-O bonds are shown by heavy lines, and the seven short Ca-O distances by heavy dashes. Megaw codes are used for Ca and T atoms, except that the operator c for a centre of symmetry is omitted.

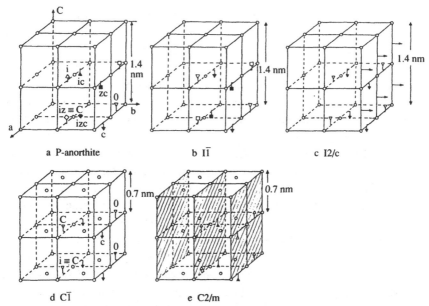

a P-anorthite   b I$\bar{1}$ c I2/c

d C$\bar{1}$ e C2/m

Fig.10.1b) Symmetry restraints on positions of M cations. Each diagram shows the M cations for a unit cell with c ≈ 1.4nm, or for two unit cells with c ≈ 0.7nm. **a** The actual positions of the four types of Ca atoms in primitive anorthite, P$\bar{1}$. The symmetry centres are shown by small circles, and relate the Ca atoms in pairs. Each pair of Ca atoms is shown by a differently shaped symbol: the absence or presence of filling merely shows the effect of the symmetry centres. The Megaw operators are used to distinguish the Ca atoms: O prototype; i body-centering; z ≈ 0.7nm translation along c; c centre of symmetry. The capital C denotes the C face-centering operation. **b** For I$\bar{1}$, there are two types of M atoms. **c** For I2/c, there is only one type of M atom, whose position is restricted by the 2-fold rotation axes shown as arrows. The c glide plane is not depicted. Note that the y-coordinate may be non-zero. **d** For C$\bar{1}$ there is only one type of M atom when c is 0.7nm. Note that the Megaw operator i is now identical to the C operator because z has become a superfluous operator. Note new centres of symmetry. **e** For C2/m, there is only one type of M atom with y-coordinate of zero to obey the mirror operation (hatched planes) (see Smith 1974).

accommodate several of these structural instabilities with the outcome that the actual crystal structure is determined by a combination of all these transition mechanisms. The interaction between the ordering schemes often leads to seemingly complicated deformation patterns of crystal structures at low temperatures. One example is the crystal structure of anorthite in Fig.10.1 which is the result of three subsequent and interacting transition mechanisms in the series C2/m - C$\bar{1}$ - I$\bar{1}$ - P$\bar{1}$ (see also Chapter 9.4). This rather complicated behaviour can, however, be greatly elucidated if we simply apply the same formal thermodynamic description to the structural properties as in the case of a phase transition with only one order parameter involved. In fact, we find that each physical mechanism leading to a real or hypothetical phase transition has to be associated with a separate order parameter and the Gibbs free energy of the system must contain terms involving not only the individual order parameters but also terms to account for the effects of interactions, or coupling, between them.

As a knowledge of the effects of coupling between various order parameters with different physical meaning is essential for the understanding of complex materials, we shall illustrate this point with an example before we proceed with the formal description. Let us return to the ferroelastic phase transition in Na-feldspar, $NaAlSi_3 O_8$, as illustrated in Fig.7.2. At high temperatures (T> 1250 K) albite is monoclinic with space group C2/m. On cooling the symmetry is reduced to triclinic ($C\bar{1}$) by two interacting processes involving a displacive lattice distortion and the ordering of Al and Si atoms between the tetrahedral sites of the feldspar framework structure. The relevant order parameter of the Al,Si ordering is called Q_{od}. The displacive transition involves the rotation of the larger lattice complexes formed by a network of tetrahedra; the so-called crankshafts (Fig.7.2). The parameter which describes this lattice distortion is called Q. Both transitions would result in the same symmetry change but the critical temperature of the displacive phase transition is 1250 K, in contrast to the lower transition temperature of the Al,Si ordering transition at ca. 983 K. Each of these transition processes could occur independently if the other transition did not occur. If, for example, the displacive transition failed to occur, the Al,Si ordering transition would still take place. As the displacive transition in Na-feldspar does occur, however, and reduces the symmetry to $C\bar{1}$ there is no further possibility for the Al,Si ordering to create another, additional phase transition and the role of the Al,Si ordering is now limited to modifications of the displacive transition mechanism. These modifications are, nevertheless, essential and it is impossible to understand the physical behaviour of Na-feldspar if either of these two transition mechanisms were ignored. We shall return to the consequences of order parameter coupling in Na-feldspar after the formal discussion of order parameter coupling in the next Chapter. Worked examples for ferroelectric/ferroelastic coupling can be found in the excellent paper by Suzuki and Ishibashi (1987) and references given there.

10.1 Coupling between two order parameters and the order parameter vector space

The coupling betweeñ spontaneous strain and the order parameter as discussed in Chapter 5 leads us conveniently to a description of two interacting order parameters. For example, if the spontaneous strain coupled with the order parameter Q_1 could itself be an order parameter Q_2, working to oppose or promote the effect of Q_1, the energy of the second order parameter Q_2 has then in itself some criticality and would lead to a phase transition if the effect of Q_1 were ignored. Let us consider the most simple case of two interacting one-dimensional order parameters both of which follow a simple Landau behaviour (extensions of the model can be treated in essentially the same way and do not contribute much to our understanding). In the case of Landau potentials with no odd order terms we may write:

$$G = \frac{1}{2} a_1 Q_1^2 + \frac{1}{4} b_1 Q_1^4 + \frac{1}{6} c_1 Q_1^6 + \frac{1}{2} a_2 Q_2^2 + \frac{1}{4} b_2 Q_2^4 + \frac{1}{6} c_2 Q_2^6$$

$$+ \lambda_1 Q_1 Q_2 + \lambda_2 Q_1^2 Q_2^2 \qquad\qquad 10.1$$

Just as for order parameter-strain coupling, the coupling terms are subject to the constraints of symmetry and cannot be introduced arbitrarily. Linear terms are excluded as they are excluded in the bare Landau potential and bilinear coupling is symmetry allowed if the transformation behaviour of $Q_1 Q_2$ is the same as Q_1^2 or Q_2^2, i.e. the product representation contains the identity representation if developed into irreducible representations. Even without going into the details of the group theoretical treatment, we can see immediately that bilinear coupling is compatible with symmetry if the symmetry properties of Q_1 and Q_2 are identical. In this case, all combinations Q_1^2, Q_2^2 and $Q_1 Q_2$ have exactly the same symmetry constraints and they are all terms of the Landau potential. The behaviour of Na-feldspar is a typical example for such a bilinear coupling. We have seen that the two order parameters Q and Q_{od} have indeed the same symmetry behaviour and would individually lead to the same change of space group during the phase transition. We can conclude, therefore, that the term QQ_{od} must also be a part of the Landau potential describing formally the coupling between Q and Q_{od}.

The second, very common coupling scheme is represented by the bi-quadratic coupling $Q_1^2 Q_2^2$. As both Q_1^2 and Q_2^2 are, by definition, part of the Landau potential we find that their product will also be symmetry allowed. This leads to the important conclusion that bi-quadratic coupling does always exist if a crystal structure relates to more than one order parameter. Uncoupled phase transitions are thus unphysical and may only appear in some approximation if the coupling energy is sufficiently small. The empirical evidence is, however, that coupling energies are normally not at all small in ferroelastic and co-elastic crystals and that bi-quadratic coupling is a very common feature (see Chapter 10.2 for a detailed coupling mechanism).

The remaining coupling energy of some importance is the linear-quadratic coupling in the form $Q_1Q_2^2$ which has to be invariant with respect to the active representation of the phase transition.

The condition that a crystal is in thermodynamic equilibrium (equation 3.2) now requires that equilibrium is reached with respect to both order parameters. The Gibbs energy is thus an energy surface spanning a two-dimensional vector space with the basis vectors Q_1 and Q_2. The equilibrium point under given thermodynamic circumstances (e.g. at a given temperature) is a total minimum of the Gibbs energy in this order parameter vector space:

$$\frac{\partial G}{\partial Q_1} = 0, \quad \frac{\partial G}{\partial Q_2} = 0$$

10.2.

This treatment can obviously be expanded to order parameter vector spaces with more than two interacting order parameters. Such situations are very common in nature (e.g. the above mentioned case of Na-feldspar) but no systematic exploration for higher dimensions than two have yet been undertaken.

We shall now discuss the general topologies of the order parameter vector spaces for the two most common coupling schemes, namely bilinear and biquadratic coupling. An equilibrium structural state which occurs under well defined temperature and pressure conditions represents one point in the order parameter vector space, as defined by the equilibrium condition 10.2. If the external conditions (e.g. T and P) are changed, we expect a different structural state to be more stable, which may or may not be related to the previous state in a continuous manner. Discontinuities can occur, however, only at points of first order phase transitions but never inside the stability field of any phase or during a continuous phase transition. A further condition for bilinear coupling is that no single order parameter can disappear without the simultaneous disappearance of the second order parameter. The obvious reason for this important rule is that there is no solution for equation 10.2. with $G_{coupling} = \lambda_1 Q_1 Q_2$ for $Q_1 \neq 0$ and $Q_2 = 0$. The only possible topology of the equilibrium line in the order parameter vector space for bilinear coupling

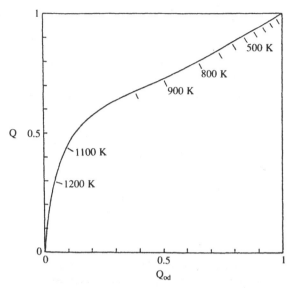

Fig.10.2 Thermodynamically stable states represented in the vector space of the order parameters Q and Q_{od}. The monalbite phase is at the origin. With decreasing temperature, the equilibrium state moves along the bent curve to $Q=1$, $Q_{od}=1$. The high albite regime is characterised by higher values of Q than Q_{od} (left part). The rounded corner represents the thermal crossover and the upper part of the curve the low albite regime (Salje et al. 1985b).

connects the origin ($Q_1=Q_2=0$, i.e. the high-symmetry phase) with the solution at absolute zero temperature which can be normalised to $Q_1=Q_2=1$. The experimental equilibrium line for Na-feldspar is shown in Fig.10.2.

Gufan and Larin (1980) were the first to point out that the intuitive assumption that similarly simple topologies also apply for higher order coupling is wrong. Salje and Devarajan (1986) derived the seven possible topologies for biquadratic coupling which are displayed in Fig.10.3. Their analytical meaning will be explained in Chapter 10.3, but the three following conclusions can be drawn from visual inspection of these diagrams:

1. The stability range of the different phases may be changed and a succession of phase transitions may occur in place of a single phase transition. Such phase transitions occur each time the equilibrium line reaches the axes of the diagrams in Fig.10.3. Cascades of phase transitions may occur if more order parameters couple. In particular, re-entrant phase transitions, where the same phase appears as a high and low temperature form, with an intermediate phase with different symmetry, may result from biquadratic coupling.

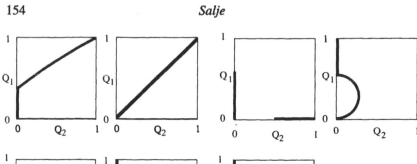

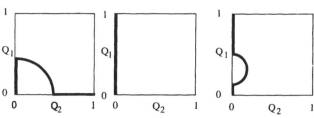

Fig.10.3 Characteristic topologies of the order parameter vector space for bi-quadratic coupling. The paraelastic phase is the origin $Q_1=Q_2=0$. Re-entrance phase transitions occur only for 2-4-6 potentials, all other topologies apply for 2-4 and 2-4-6 potentials.

2. The order of a phase transition is influenced by the coupling. As in the case of the order parameter - strain coupling in Chapter 5, we find that a continuous phase transition can become discontinuous and vice versa due the coupling energy.

3. The temperature evolution of the order parameter(s) near the transition point can be completely different from the expected simple Landau behaviour of one single order parameter. This does not, in general, indicate that the Landau approach is wrong but simply means that the interaction between order parameters has not been appreciated.

10.2 Coupling via common strain components, a possible coupling mechanism

The physical origin of order parameter coupling can vary depending on the structural properties of the material under investigation. As long as we are interested in the thermodynamic description of the material, the different coupling mechanisms do not really matter because they will all be adequately described by the Landau potential. The details of the phase transition process are still universally related to the order parameters and the structural behaviour depends directly upon them. On the other hand, if we try to give physical meaning to the various parameters in the Landau potential itself, we can gain confidence in the reliability of the coupling energies as determined indirectly from a thermodynamic treatment. It is, for example, normally expected that coupling energies describing the interaction between a spin-related order parameter and an order parameter correlated with a structural distortion will be small because we know that spin-phonon interactions contain relatively small amounts of energies (compared with phonon-phonon interactions). Cation ordering, on the other hand, will presumably couple strongly with lattice distortions, in particular if the cations involved have different sizes.

One specific coupling mechanism which appears to be a common feature of ferroelastic and co-elastic materials is based on the elastic interactions between the two order parameters. Following the treatment of the order parameter-strain coupling in Chapter 5 we know that each order parameter will create a spontaneous strain e_S. Let us now assume that the spontaneous strain created by either of the two order parameters has at least one common component e. We find that the Gibbs free energy:

$$G(Q_1,Q_2,e) = \frac{1}{2} a_1 Q_1^2 + \frac{1}{4} b_1 Q_1^4 + \frac{1}{6} c_1 Q_1^6 +$$

$$+ \frac{1}{2} a_2 Q_2^2 + \frac{1}{4} b_2 Q_2^4 + \frac{1}{4} c_2 Q_2^6$$

$$+ d_1 Q_1 e + d_2 Q_2 e + \varepsilon_1 Q_1^2 e + \varepsilon_2 Q_2^2 e + f e^2$$
$$10.3$$

implicitly contains the coupling between the two order parameters. This coupling becomes obvious if we apply the condition that the crystal is free of stress:

$$\frac{\partial G}{\partial e} = 0$$
$$10.4$$

and find for $\varepsilon_1 = \varepsilon_2 = 0$

$$e = - \frac{1}{2f} (d_1 Q_1 + d_2 Q_2)$$
$$10.5$$

The Gibbs free energy is then:

$$G = \frac{1}{2}\left(a_1 - \frac{d_1^2}{2f}\right)Q_1^2 + \frac{1}{4}b_1\,Q_1^4 + \frac{1}{6}c_1\,Q_1^6$$

$$+ \frac{1}{2}\left(a_2 - \frac{d_2^2}{2f}\right)Q_2^2 + \frac{1}{4}b_2\,Q_2^4 + \frac{1}{6}c_2\,Q_2^6 - \frac{d_1 d_2}{2f}\,Q_1\,Q_2 \qquad 10.6$$

This Landau potential is now identical with the original Gibbs free energy in equation 10.3 where the coupling terms are directly related to the elastic constants and to the magnitude of the spontaneous strain via equation 10.5. for the bilinear case and via:

$$e = -\frac{1}{2f}\left(\varepsilon_1\,Q_1^2 + \varepsilon_2\,Q_1^2\right) \qquad 10.7.$$

for the biquadratic coupling.

The correlation between the spontaneous strain, the elastic constants and the effective coupling constants show clearly that a soft material with a large spontaneous strain will always be characterized by a strong coupling between the two order parameters, whereas a hard material with a small lattice distortion due to the phase transition is less susceptible to this coupling mechanism. As the elastic constants and the relevant strain components are normally known from experimental observations, it is always possible to estimate the coupling strength using equation 10.6. Let us consider a typical example for a crystal with bilinear order parameter coupling for the most simple case that the strength of the coupling between the order parameters and the strain is identical for both order parameters. The numerical value of the coupling constant between the two order parameters is then identical to half the elastic energy at absolute zero (i.e. $Q_1 = Q_2 = 1$). Experimental values in Na feldspar lead to an elastic energy at zero temperature of ca.-6 kJ/mol , those in $NaNO_3$ are around -4 kJ/mol. Typical coupling energies are half these values and can be expected to reach some 3 kJ/mol. The total Gibbs free energies in these ferroelastic and co-elastic phase transitions are around -10 kJ/mol. The energy involved in the coupling between two order parameters due to a common strain component can thus be as large as some 30% of the total excess Gibbs free energy in co-elastic materials, a proportion which clearly demonstrates the importance of this specific coupling mechanism.

10.3 Analytical solutions for bilinear coupling

In the simplest treatment of a coupling phenomenon it is assumed that the coupling process can be described entirely by the bilinear terms in equation 10.3. The excess Gibbs free energy can then be written for the most simple Landau potential as:

$$G = \frac{1}{2} A_1 \left(T\text{-}T_{c1}^*\right) Q_1^2 + \frac{1}{4} b_1 Q_1^4 + \frac{1}{6} c_1 Q_1^6$$

$$+ \frac{1}{2} A_2 \left(T\text{-}T_{c2}^*\right) Q_2^2 + \frac{1}{4} b_2 Q_2^4 + \frac{1}{6} c_2 Q_2^6$$

$$+ \lambda_1 Q_1 Q_2 \qquad\qquad 10.8$$

with:

$$T_{c1}^* = T_{c1} + \frac{d_1^2}{2fA_1} \;,\; T_{c2}^* = T_{c2} + \frac{d_2^2}{2fA_2} \qquad\qquad 10.9$$

and:

$$\lambda_1 = -\frac{d_1 d_2}{2f} \qquad\qquad 10.10$$

The condition that the system is in thermodynamic equilibrium for any combination of order parameters is:

$$\frac{\partial G}{\partial Q_1} = \frac{\partial G}{\partial Q_2} = 0 \qquad\qquad 10.11$$

which leads to two basic equations which have to be satisfied simultaneously:

$$a_1^* Q_1 + b_1 Q_1^3 + c_1 Q_1^5 + \lambda_1 Q_2 = 0 \qquad\qquad 10.12$$

$$a_2^* Q_2 + b_2 Q_2^3 + c_2 Q_2^5 + \lambda_1 Q_1 = 0 \qquad\qquad 10.13$$

with:

$$a_1^* = a_1 - \frac{d_1^2}{2f} = A_1 \left(T\text{-}T_{c1}^*\right) \qquad\qquad 10.14$$

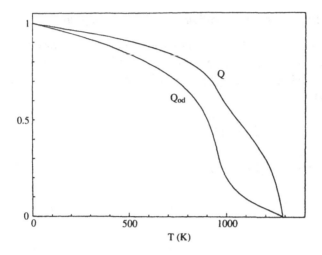

Fig.10.4 Temperature dependence of the 'displacive' order parameter Q and the Al,Si-order parameter Q_{od} at equilibrium.

$$a_2^* = a_2 - \frac{d_2^2}{2f} = A_2\left(T - T_{c2}^*\right)$$ 10.15

with the renormalized temperatures T_{c1}^* and T_{c2}^* given in equation 10.9.

Only two solutions of the equations 10.12. and 10.13 exist which define the two thermodynamic phases:

phase O : $Q_1 = 0$, $Q_2 = 0$ 10.16

phase I : $Q_1 \neq 0$, $Q_2 \neq 0$ 10.17

Note that no other phase is allowed which is in agreement with the topology of the equilibrium line in the order parameter vector space as discussed in 10.2.

The effect of the coupling is thus the renormalization of the two temperatures T_{c1} and T_{c2}. If the higher of the two critical temperatures is related to a second order phase transition we find that the actual phase transition occurs at this renormalized temperature. The lower of the two temperatures is no longer associated with a phase transition; instead it leads to a second anomaly of the two order parameters at this point. This point is thus called a 'crossover point' between two thermal regimes, with the same symmetry but with different ratios between the two order parameters. Most physical properties of the crystal, such as the specific heat, the elastic constants etc., change quite dramatically at the crossover point, and these changes might resemble the changes normally observed during a structural phase transition. It is easy, therefore, to mistake a crossover point for a transition point if no further structural information is available. In Fig. 9.11 the anomalies in the thermodynamic properties of Na-feldspar are illustrated where crossover behaviour is observed near 900K. This crossover separates the thermal regime of low Al,Si order (Q_{od}) and large lattice distortions (Q) (often called 'high albite') from the low temperature regime with large degrees of Al,Si order (Q_{od}) and large lattice distortions (Q), often called 'low albite'.

An important feature of the bilinear coupling is that the crossover behaviour is rather smooth even if the original, uncoupled phase transition would have been stepwise. The physical reason behind this smoothing effect is the role played by the order parameter which dominates at higher temperatures and which acts as a conjugate field for the second order parameter (see Chapter 9.3). As all positive conjugated fields stabilise the low-symmetry phase we find that the stepwise behaviour of Q_{od} (for the hypothetical case of Q=0) becomes continuous and that, with increasing temperature Q_{od} can only disappear when Q disappears. This behaviour is illustrated in Fig.10.4 where both order parameters are plotted individually as functions of temperature.

The energetics of the coupling process in Na-feldspar have been determined experimentally (Salje et al. 1985b). Using their experimental values of the coefficients in equations 10.12 and 10.13 we can now write two basic equations describing the thermodynamic equilibrium in a numerical form:

$$5.479 \, (T - 1251) \, Q + 6854 \, Q^3 + d \, Q_{od} = 0$$

$$41.62 \, (T - 824) \, Q_{od} - 9301 \, Q_{od}^3 + 43600 \, Q_{od}^5 + dQ = 0$$

$$d = - 2.171 - 3.043 \, T - 0.0016 \, T^2 + 2.1 \cdot 10^{-6} \, T^3 \qquad\qquad 10.18$$

where all the energies are in units of J/mol and all temperatures are in K.

Note here that the coupling energy for $T \sim 1000K$ is of the same order of magnitude as the enthalpy of the displacive phase transition and that the coupling process plays thus an essential part in the phase transition mechanism.

Finally, we give below an analytical expression for the solution of equations 10.12 and 10.13 in terms of only one order parameter:

$$\frac{c_1c_2}{\lambda_1^5}Q_1^{24} + \frac{5b_1c_1^4c_2}{\lambda_1^5}Q_1^{22} + \left(\frac{5a_1^*c_1^4c_2}{\lambda_1^5} + \frac{10b_1^2c_1^3c_2}{\lambda_1^5}\right)Q_1^{20}$$

$$+\left(\frac{10b_1^3c_1^2c_2 + 20a_1^*b_1c_1^3c_2}{\lambda_1^5}\right)Q_1^{18}$$

$$+\left(\frac{5b_1^4c_1c_2 + 10a_1^2c_1^3c_2 + 30a_1^*b_1^2c_1^2c_2}{\lambda_1^5}\right)Q_1^{16}$$

$$+\left(\frac{b_1^5c_2 + 30a_1^2b_1c_1^2c_2 + 20a_1^*b_1^3c_1c_2}{\lambda_1^5} + \frac{c_1^3b_2}{\lambda_1^3}\right)Q_1^{14}$$

$$+\left(\frac{5a_1^*b_1^4c_2 + 10a_1^3c_1^2c_2 + 30a_1^2b_1^2c_1c_2}{\lambda_1^5} + \frac{3b_1b_2}{\lambda_1^3}\right)Q_1^{12}$$

$$+\left(\frac{10a_1^2b_1^3c_2 + 20a_1^3b_1c_1c_2}{\lambda_1^5}\; \frac{3a_1^*c_1^2b_2 + 3b_1^2b_2c_1}{\lambda_1^3}\right)Q_1^{10}$$

$$+\left(\frac{5a_1^4c_1c_2 + 10a_1^3b_1^2c_2}{\lambda_1^5}\; \frac{b_1^3b_2 + 6a_1^*b_1b_2c_1}{\lambda_1^3}\right)Q_1^8$$

$$+\left(\frac{5a_1^4b_1c_2}{\lambda_1^5} + \frac{3a_1^*b_1^2b_2 + 3a_1^2b_2c_1}{\lambda_1^3}\right)Q_1^6$$

$$+\left(\frac{a_1^5c_2}{\lambda_1^5} + \frac{3a_1^2b_1b_2}{\lambda_1^3} + \frac{c_1a_2^*}{\lambda_1}\right)Q_1^4 + \left(\frac{a_1^3b_2}{\lambda_1^3} + \frac{b_1a_2^*}{\lambda_1}\right)Q_1^2$$

$$+\left(\frac{a_1^*a_2^*}{\lambda_1} - \lambda_1\right) = 0 \tag{10.19}$$

From this polynomial, it can be seen that a direct determination of G in this form is virtually impossible, whereas the coefficients in the equivalent equations 10.12 and 10.13 can easily be found experimentally. The effect of coupling, even in the most simple bilinear case, is thus to dramatically increase the order of terms in the Landau potential, which would be a rather unphysical effect for a single order parameter. One might suspect, therefore, that in cases where experimental findings in co-elastic crystals cannot be successfully described within the framework of a Landau potential with terms up to sixth order, that in these cases additional order parameter coupling occurs which has not been correctly accounted for.

10.4 Biquadratic coupling between two order parameters

Biquadratic coupling is always allowed by symmetry and it will link order parameters even if they have different symmetry properties. Typical examples are the coupling between order parameters which act at different points at the boundary of the Brillouin zone in the Palmierite structure (e.g. Pb_3 $(AsO_4)_2$, Bismayer et al. 1986), in the case of the influence of Al,Si order on the phase transition $P\bar{1} - I\bar{1}$ in anorthite (Salje 1987, Redfern et al. 1987) or the postulated coupling in $NaNO_3$ (Schmahl and Salje 1989). The Gibbs free energy for strain induced coupling is:

$$G = \frac{1}{2} a_1 Q_1^2 + \frac{1}{4} b_1 Q_1^4 + \frac{1}{6} c_1 Q_1^6 + \frac{1}{2} a_2 Q_2^2 + \frac{1}{4} b_2 Q_2^4 + \frac{1}{6} c_2 Q_2^6$$

$$+ \varepsilon_1 Q_1^2 e + \varepsilon_2 Q_2^2 e + fe^2 \qquad\qquad 10.20$$

The stress free condition yields:

$$e = -\frac{1}{2f} \left(\varepsilon_1 Q_1^2 + \varepsilon_2 Q_2^2 \right) \qquad\qquad 10.21$$

and the effective Landau potential becomes:

$$G = \frac{1}{2} a_1 a_1^2 + \frac{1}{4} \left(b_1 - \frac{\varepsilon_1^2}{f} \right) Q_1^4 + \frac{1}{6} c_1 Q_1^6$$

$$+ \frac{1}{2} a_2 Q_2^2 + \frac{1}{4} \left(b_2 - \frac{\varepsilon_2^2}{f} \right) Q_2^4 + \frac{1}{6} c_2 Q_2^6$$

$$- \frac{\varepsilon_1 \varepsilon_2}{2f} Q_1^2 Q_2^2 \qquad\qquad 10.22$$

We see that, in contrast to bilinear coupling, it is the fourth order term which is renormalised by the strain coupling. The same result was found for coupling between the strain and one single order parameter in Chapter 5.2. The effective coupling parameter between the two order parameters is now:

$$\lambda_2 = -\frac{\varepsilon_1 \varepsilon_2}{2f}$$
10.23

The sign of the coupling constant is negative if the two order parameters promote each other, i.e. their combination lowers the Gibbs free energy. The sign is positive if the order parameters are in competition. There are four possible phases which represent mimina of the Gibbs free energy:

Phase O with $Q_1 = Q_2 = 0$
10.24

This is the high-symmetry phase.

Phase I with $Q_1 = 0$

$$Q_2^2 = \frac{1}{2c_2}\left[-b_2^* \pm \sqrt{b_2^{*2} - 4a_2 c_2}\right]$$
10.25

Phase II with $Q_1^2 = \frac{1}{2c_1}\left[-b_1^* \pm \sqrt{b_1^{*2} - 4a_1 c_1}\right]$

$$Q_2 = 0$$
10.26

Phase III with $Q_1 \neq 0$, $Q_2 \neq 0$

Phase III represents a new structural state which would not occur without order parameter coupling. The order parameter Q_1 is given implicitly by the condition:

$$4\lambda_2^2 a_2 - 2\lambda_2 b_2^* a_1 + c_2 a_1^2 + \left[-2\lambda_2 b_1^* b_2^* + 2a_1 c_2 b_1^* + 8\lambda_2^3\right]Q_1^2$$

$$+\left[-2\lambda_2 b_2^* c_1 + c_2 b_1^{*2} + 2a_1 c_1 c_2\right]Q_1^4 + 2c_1 c_2 b_1^* Q_1^6 + c_1^2 c_2 Q_1^8 = 0 \quad 10.27$$

The order parameter Q_2 follows from exactly the same equation when the indices 1 and 2 are interchanged. In this equation we have used the renormalised coefficients of the fourth order terms:

$$b_1^* = b_1 - \frac{\varepsilon_1^2}{f}$$
10.28

$$b_2^* = b_2 - \frac{\varepsilon_2^2}{f}$$
10.29

Comparing the results of the bilinear coupling (where only two phases occur) with the four possible stable configurations in the case of biquadratic coupling, makes us realize that non-linear coupling leads to a wealth of transitions between phases with distinctly different structural features. The sequence of phase transitions, such as displayed in the order parameter vector space in Chapter 10.1 depends on the numerical values of the coefficients of the Landau potentials. Each sequence follows one of the seven topologies of the equilibrium lines in Fig.10.3. The topology of the equilibrium lines is largely controlled by the numerical values of the coupling constants: details of the various parameter combinations have been discussed and compared with experimental results by Salje and Devarajan (1986), Bismayer et al. (1986) and Schmahl and Salje (1989).

The analytical treatment of the coupling phenomena becomes simpler if we assume that one of the two bare phase transitions is second order (i.e. $c_2=0$, $b_2 > 0$). The temperature evolution of the order parameters Q_1 and Q_2 can then be written explicitly as:

$$Q_1^2 = \frac{1}{b_2^* c_1} [-\frac{1}{2} b_1^* b_2^* + 2 \lambda_2^2 \pm (4 \lambda_2^4 + \frac{1}{4} b_1^{*2} b_2^{*2} - 2 \lambda_2^2 b_1^* b_2^*$$

$$+ 2 \lambda_2 b_2^* c_1 a_2 - a_1 c_1 b_2^{*2})^{\frac{1}{2}}]$$

10.30

$$Q_2^2 = \frac{1}{b_2^{*2} c_1} [\lambda_2 b_1^* b_2^* - a_2 c_1 b_2^* - 4 \lambda_2^3 \pm ((4 \lambda_2^3 - \lambda_2 b_1^* b_2^*)^2$$

$$+ 4 \lambda_2^2 b_2^* c_1 (2 \lambda_2 a_2 - a_1 b_2^*))^{\frac{1}{2}}]$$

10.31

If both sixth order terms vanish, we find that the solution becomes trivial with:

$$Q_1^2 = \frac{a_1 b_2^* - 2 \lambda_2 a_2}{4 \lambda_2^2 - b_1^* b_2^*}$$

10.32

$$Q_2^2 = \frac{a_2 b_1^* - 2 \lambda_2 a_1}{4 \lambda_2^2 - b_1^* b_2^*}$$

10.33

Under these conditions all the phase transitions which are not topological stepwise are second order.

11

GRADIENT COUPLING AND STRAIN MODULATIONS

So far crystals were assumed to be homogeneous; now we allow the order
parameter to vary as a function of space and time. This may cost energy (i.e. the
Ginzburg energy) and release solitary waves. Modulated phases, precursor
order and embryonic states are treated as specific signatures of solitary states.

We have seen in the last chapter that the coupling between two order
parameters is the origin of a multitude of new and rather unexpected physical
phenomena which do not occur in a Landau-Ginzburg system with only one
order parameter. The coupling discussed so far was related to the interaction
between two homogeneous order parameters both of which were analysed in
terms of individual Landau potentials and an additional coupling term. We
now make the same extension of Landau theory as in Chapter 8 and allow the
order parameter to vary in space. The length scale is again mesoscopic to
ensure that a thermodynamic treatment is justified to a good approximation.
Following the arguments in Chapter 8 we find that the additional energy
required for the appropriate description of the variation of Q(r) is the
Ginzburg energy which we write as a function of the gradient of the order
parameter:

$$G_{Ginzburg} = \frac{1}{2} g \left(\nabla Q\right)^2 + \frac{1}{2} g' \left(\nabla^2 Q\right)^2 + \frac{1}{4} g'' \left(\nabla Q\right)^4 + \dots \qquad 11.1$$

Although there is evidence that gradient energies of the order four and
higher are relevant, e.g. for an understanding of the behaviour of martensites,
very little work has yet been undertaken on their physical and mathematical
analysis. Here we limit ourselves for most of the discussion to the harmonic
limit (i.e. g'=g''=0).

The Ginzburg energy in equation 11.1 can be envisaged as a Landau
potential of the gradient of the order parameter rather than the order
parameter itself. In other words, the energy resource for the phase transition is
related to the amount of energy necessary to bend the order parameter in the

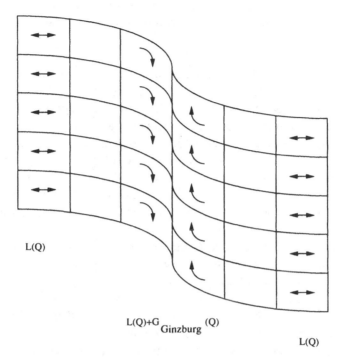

Fig.11.1 Two dimensional plot of bent lattice planes indicating the various energy contributions. The lattice on either end of the drawing is distorted due to the structural phase transition. The spontaneous strain is symbolised by the double arrows. The relevant energy is the Landau potential. When the planes are bent, additional gradient energy occurs as described by the Ginzburg energy.

structure, very similar to the elastic energy related to warping of lattice planes in alloys. In Fig.11.1 a simple model is used to illustrate the two energy distributions. In this model we see that the Landau potential causes the atoms to be displaced in the low-symmetry phase. The Ginzburg energy leads to spatial variations in the form of solitary waves.

If the Ginzburg energy is positive, the formation of this kink leads to a less stable configuration and thus to a higher Gibbs free energy. If, on the other hand, the Ginzburg energy is negative, the total Gibbs free energy is reduced by each kink and the crystal will tend to nucleate as many kinks as possible in order to increase its thermodynamic stability. The density of kinks in the structure is then limited by the kink-kink repulsion leading to the formation of kink lattices often referred to as soliton lattices. As the relevant forces for the formation of these soliton lattices are the energy gain per soliton wall and the inter-wall repulsion, we see that in the first approximation, this soliton lattice is independent of the actual crystal structure and, in particular, is independent of the periodicity of the underlying crystal lattice. There is thus no reason why the repetition unit of the solitons should relate to the crystallographic lattice parameters. If these two lattices do not correlate we call the structural phase an **incommensurate phase**. In many cases, there are, however, weak interactions between the two lattices and an energetically favorable situation is reached when there is some match between the two periodicities in which case the phase is called a **'lock-in' phase**. Both the incommensurate and the lock-in phases are characterised by the presence of structural modulations due to a superimposed soliton lattice and both types of crystal structures are classified under the more general term of **'modulated phases'**. In Fig.11.2, examples of an incommensurate phase derived from computer simulations are shown.

Before we discuss structural features of the soliton lattice, let us illustrate the relevance of the Ginzburg energy to a physical situation in which the modulated phase has already been formed. In ferroelastic and co-elastic crystals we expect the structural order parameter to be coupled with spontaneous strain. The gradient of the order parameter then has the meaning of bending the lattice planes as in Fig.11.1, normally as a shear distortion. If this bending distortion occurs periodically, we can describe the order parameter as a periodic function where the periodicity of the soliton lattice defines its wavelength. The periodicity of the underlying lattice is already part of the homogeneous order parameter and does not appear in this formalism explicitly. The structural distortion can therefore be described by the wave functions:

$$Q_1^* = Q_1 \exp(ikr) = Q_0 \exp(i[kr + f])$$
$$Q_2^* = Q_2 \exp(-ikr) = Q_0 \exp(i[kr + f])$$

$$11.2$$

where Q_0 is the amplitude of the distorted wave, k is the wavevector, r is the space vector along the propagation direction of the wave and ϕ is the phase angle between the two waves. A homogeneous order parameter Q_0 would occur for an infinite wavelength, i.e. k=0 and vanishing phase angle ϕ, with $Q_1 = Q_2 = Q_0$. As the crystal structure of the high-symmetry phase is periodic with reciprocal lattice vectors a*, b* and c* any k-vector which corresponds to a lattice point in the reciprocal lattice will also fulfill this invariance condition. We shall return to the role of such lattice points in the next chapter and restrict ourselves here to the case k=0. The Gibbs free energy can now be

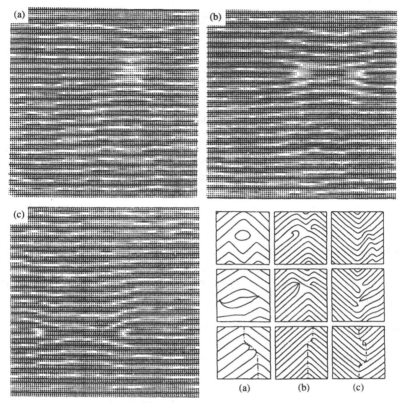

Fig.11.2a) Examples of incommensurate modulations as obtained by molecular dynamics simulaton. The wavevectors are 0.375 (a), 0.385 (b) and 0.395 (c). Dynamical variations of the wavevectors occur via local defects where several links merge (so-called stripples) (after Parlinski 1988).
b) Time evolution of kinks between areas with irregular soliton lattices. The time runs from top to bottom of the drawing for the three configurations a,b,c. The 'final' states display the typical herringbone pattern often observed in modulated crystals (after Parlinski 1988).

constructed in the same way as shown in Chapter 10 with the additional condition that the Gibbs free energy must be invariant with respect to translations of the crystal as a whole. This invariance is now imposed on each term of the Landau-Ginzburg polynomial, leading to the condition that the total wavevector has to be conserved. This means in our present example that all phase factors have to disappear; in particular the sum of all k-vectors has to be zero. The excess Gibbs free energy as a function of Q_1 and Q_2 can then be written in the form:

Fig.11.2c) Microstructure of quartz in an inhomogeneous temperature field. The lowest temperature is at the top-right hand side showing coarse Dauphine twins which are characteristic of the low symmetry form. The regular pattern of bent rows of triangles at the lower left hand side displays a structural modulation which matches the geometrical features of the incommensurate phase. The dark area at the bottom left hand side is the hottest spot of the crystal where no structural modulations can be observed. This structural state represents on a mesoscopic scale the paraelastic phase (Courtesy of Van Tendeloo and Van Landuyt).

$$G(Q_1,Q_2) = \frac{1}{2} A(T\text{-}T_c)\, Q_1 Q_2 + \frac{1}{4} B\, Q_1^2 Q_2^2 + ... + \frac{1}{2}\, g\, k^2\, Q_1 Q_2 \qquad\qquad 11.3$$

Note that the conservation law for the wavevector is incompatible with any term of the classical Landau potential (e.g. Q_1^2 or Q_1^4) and that equation 11.3 contains only the coupling terms introduced in Chapter 10. We can now substitute the order parameters Q_1 and Q_2 by their common amplitude (equation 11.2) and find:

$$G(Q_o) = \frac{1}{2}\left[A(T\text{-}T_c) + g\, k^2 \right] Q_o^2 + \frac{1}{4} B\, Q_o^4 + ... \qquad\qquad 11.4$$

We see that a structural phase transition is now expected to take place at the renormalised temperature $T_c^* = T_c - gk^2/A$ which is higher than T_c for negative values of the gradient term g. The critical wavevector of the modulation is $k=k_m$ which is normally a function of temperature and pressure.

It is now useful to study the stability of the structural state in the modulated phase with respect to the two possible degrees of freedom. Firstly, we may allow the amplitude of the order parameter to fluctuate. The susceptibility of the system is then (see equation 5.15):

$$\chi_L = \left(\frac{\partial^2 G}{\partial Q_0^2}\right)^{-1} = [A(T-T_c^*)]^{-1} \text{ at } T > T_c \qquad\qquad 11.5$$

$$= \frac{1}{2}[A(T-T_c^*)]^{-1} \text{ at } T < T_c \qquad\qquad 11.6$$

The amplitude fluctuations thus have the same stability as in the case of a non-modulated phase transition. These fluctuations are generally called **amplitudons,** a term first introduced by Overhauser (1971), the index L in χ_L indicates the longitudinal character of the fluctuations with respect to the amplitude of the modulation.

We now consider the susceptibility of the systems with respect to phase changes χ_T. The index T indicates that the movements during the fluctuations are now transverse (or orthogonal) with respect to the amplitude of the modulation. Let us consider variations of the phase of the wavevector near the critical value k_m as:

$$k = k_m + \Delta k \qquad\qquad 11.7$$

We can now define a fluctuation of the phase via the equivalent transverse susceptibility. In harmonic approximation:

$$\chi_T^{-1} = \frac{\partial^2 G}{\partial \phi^2} \propto k^2 \qquad\qquad 11.8$$

describing the phase change of the modulation during the fluctuation. The equivalent lattice particle is called a **'phason'**. The phason dispersion does not explicitly depend on temperature or pressure although the gradient term g might implicitly contain such dependencies. There is, however, no criticality of χ_T (in contrast to χ_L) and we find that the frequency of the phason always disappears for $k=k_m$ (i.e. $\Delta k=0$) indicating that the modulation can be phase shifted through the crystal without having to overcome restoring forces. The dispersion relations for the amplitudons and phasons are depicted in Fig.11.3. Figure 11.4 shows experimental results for the temperature evolution of the amplitudon frequency in $ThBr_4$ and ferroelastic $Pb_3(PO_4)_2$.

We can now return to a discussion of the nature of the order parameter-strain coupling. The symmetry requirement of the invariance of the

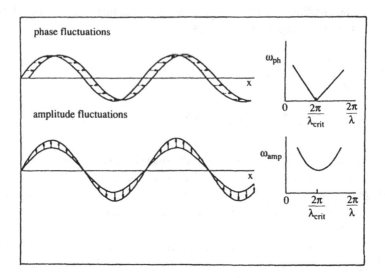

Fig.11.3a) Sketch of the phase (top) and amplitude (bottom) fluctuations in an incommensurate phase (plane wave limit). On the right hand side, the equivalent dispersion curves are shown.

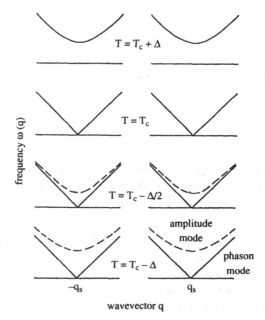

Fig.11.3b) The dispersion relations for the excitation frequencies of an incommensurate system with a plane-wave ground state. At $T<T_c$ the amplitudon and phason dispersions are shown (after Bruce and Cowley 1981).

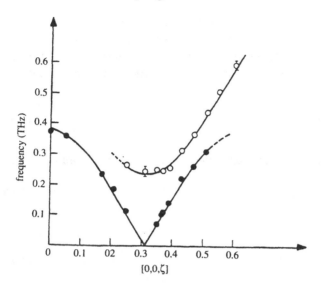

Fig.11.4a) Phase (black circles) and amplitude (open circles) dispersion in ThBr$_4$ (Currat et al.,1986).

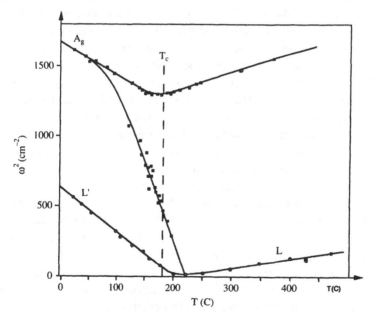

Fig.11.4b) Temperature dependence of phonon frequencies in Pb$_3$(PO$_4$)$_2$, (squares: Raman scattering, circles: Neutron scattering). The ferroelastic phase transition takes place at T$_c$=180°C. Between T$_c$ and ca.260°C, low symmetry clusters with dynamical walls similar to random incommensurations occur. These clusters do not generate macroscopic spontaneous strain.

wavevector leads to the simple conclusion that no energy contribution can exist which is linear in the order parameter or quadratic in one order parameter. As the spontaneous strain is a homogeneous quantity (i.e. k=0), the lowest order of coupling with order parameters in a modulated structure is quadratic-linear:

$$G_{coupling} = Q_1 Q_2 e \qquad\qquad 11.9$$

which, when transformed into amplitudon and phason coordinates becomes:

$$G_{coupling} \text{ (amplitudon)} = Q_0^2 e \qquad\qquad 11.10$$

and

$$G_{coupling} \text{ (phason)} = 0 \qquad\qquad 11.11$$

Consequently the strain coupling in this approximation alters the amplitude of the modulation wave but not the phase. As the quadratic-linear coupling is normally smaller than the bi-linear coupling, we find also that the magnitude of the spontaneous strain as built up in the modulated phase is significantly smaller (often by one order of magnitude) than in a co-elastic low-symmetry phase with bi-linear coupling between the order parameter and the strain. In Fig.11.5 we see a typical situation for the formation of spontaneous strain as observed in plagioclase feldspars. The spontaneous strain is largest for the Na-rich plagioclases with bi-linear coupling in the case of Na-feldspar (see Chapter 11.4). The Ca-rich plagioclases show the same coupling for the phase transition $C2/m$-$C\overline{1}$ which is then superceded by the second phase transition $C\overline{1}$ - $I\overline{1}$. The total spontaneous strain with respect to the monoclinic phase has the same order of magnitude as in Na-feldspar but with the opposite sign. Modulated plagioclases (so-called 'e-plagioclases') occur in the middle of the albite-anorthite phase field (Ab-An). No bi-linear coupling is symmetry allowed for e-plagioclases and the resulting spontaneous strain is thus much smaller than for the pure albite and anorthite components. This does not, however, mean that the strain amplitude associated with the modulation is small; it only indicates that the spontaneous strain as observed macroscopically is insensitive to these local modulations.

As the amplitude of the lattice distortion Q_0 is a close image of the local strain amplitude, we can envisage that the bi-linear coupling disappears in the modulated phase and thus reduces the spontaneous strain significantly as a self-compensation of the structural strain. This self-compensation reduces the

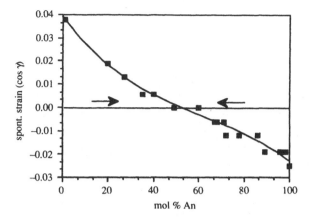

Fig.11.5 Composition dependence of the spontaneous strain in plagioclase feldspar. Structurally modulated states occur near the intersection of the curve with the strain-free line. Experimentally, modulated structures were observed in the composition range between the two arrows (after Carpenter et al. 1985).

elastic energy and leads to stabilisation of the modulated phase. It is an empirical observation, however, that this self-compensating role of the strain in a modulated phase is often limited to small amplitudes (e.g. $e_{local}<0.01$). The physical reason for this behaviour lies in the non-harmonic part of the gradient energy in equation 11.1 which increases the Gibbs free energy of the modulated phase with the fourth power of the modulation amplitude which exceeds the stabilising role of the negative quadratic gradient energy for large amplitudes. Let us call the renormalized fourth order coefficient B^*. The Gibbs free energy for the modulated phase in thermodynamic equilibrium for a simple second order phase transition is then:

$$G_{mod} = -\frac{1}{4}\left(A^2/B^*\right)(T-T_c^*)^2 \qquad\qquad 11.12$$

as compared with the non-modulated low-symmetry phase, including the strain energy in the term:

$$G_o = -\frac{1}{4}\left(A'^2/B\right)(T-T_c)^2 \qquad\qquad 11.13$$

The modulated phase becomes unstable at a temperature T_0 at which both Gibbs free energies are equal:

$$\left(\frac{T_o\text{-}T_c}{T_o\text{-}T_c^*}\right)^2 = \frac{B}{B^*}\left[\frac{A}{A'}\right]^2$$

11.14.

Let us assume values $T_c^*=1.1T_c$ and $T_0=0.9T_c$. If we further allow for a maximum spontaneous strain in the non-modulated phase of $e=0.01$ we find a maximum local strain in the modulated phase $e_{local}= 0.001$ which is the typical order of magnitude of experimentally observed strain parameters. Note that larger strain amplitudes such as observed in e-plagioclases are often indicative of smaller values of T_0 which is the lower temperature limit of the stability of the modulated phase.

11.1 Lock-in phase transitions

So far we have discussed structural modulations within the framework of the Landau-Ginzburg approach ignoring the role of the underlying crystal lattice. We have found that the basic features of the amplitudon and phason dispersions describe the physical behaviour of the crystal rather well and we have identified the strain modulation as one prominent reason for the structural modulation that occurs in co-elastic crystals. None of these arguments, however, could give us detailed information about the wavelength of the modulation beside the rather general argument that this wavelength is determined by the minimum of the amplitudon dispersion curve. In fact, we find that in a large number of incommensurate crystals the wavelength changes as a function of temperature, pressure or external fields. This variation can appear to be rather continuous (Fig.11.6) although in most cases closer inspection reveals a discontinuous behaviour (Fig.11.7). In the latter case, small steps with commensurate wavevectors are separated by intervals with truly incommensurate modulations. The stepwise behaviour in the k *versus* T curves is called a **'devil's staircase'** (Fig.11.7) which has been extensively studied using the microscopic Frenkel-Kontorova one-dimensional model (e.g. Aubry 1983). As introduced in the last chapter, each structural state with a constant commensurate wavevector k_m over a finite temperature interval has the character of a thermodynamic phase in its own right. Such phases are called 'lock-in' phases if they occur as part of a devil's staircase. It was found that incommensurations between the steps in the devil's staircase occur at temperatures close to the transition point which separates the high symmetry phase from the modulated phase, whereas no such steps occur at lower temperatures close to the homogeneous low-symmetry phase. There is also a strong indication that chaotic states occur

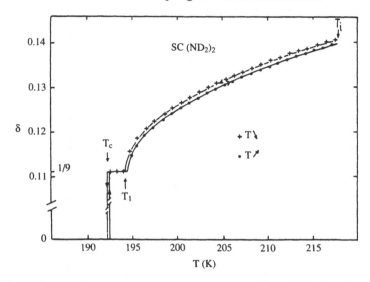

Fig.11.6 Example for a continuous temperature dependence of the wavelength of the modulation in $SC(ND_2)_2$. A lock-in phase occurs for $\delta=1/9$ (after Durand et al. 1985).

before the structure becomes locked at one of the most prominent lock-in phases (e.g. Blinc et al. 1983).

Lock-in phases occur because of correlations between the crystal structure and its periodicity and the periodicity of the structural modulations - so far ignored in our treatment of the Landau-Ginzburg theory. It is clear that such correlations emerge 'automatically' from atomistic models but rather than using this coup de force it appears that there is also an elegant thermodynamic argument which allows us to gain some insight into the nature of lock-in phases. Let us return to the conservation law of the k-vectors of the interacting order parameters. In Chapter 11 we have restricted ourselves to the case where the sum of the k-vectors is zero:

$$\Sigma\, k_i = 0 \qquad\qquad\qquad 11.15$$

where we sum over all amplitudes occurring in the interaction.

We now include the symmetry equivalent points of the reciprocal lattice t*:

$$\Sigma\, k_i = t^* \qquad\qquad\qquad 11.16$$

where t* is a reciprocal lattice vector. We can now extend the Landau-Ginzburg polynomial to include all those terms which fulfill the condition

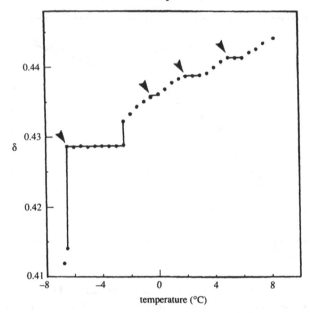

Fig.11.7 Temperature dependence of the modulation wavevector in [N(CH$_3$)$_4$]$_2$
FeCl$_4$ in a cooling experiment. Steps on the devil's staircase are indicated by
horizontal lines. The low stability temperatures are shown by arrowheads (data from
Mashiyama and Tanisaki, 1982).

11.16. These terms are called the 'Umklapp invariants' in contrast to the
'normal invariants' in 11.15. The Umklapp energies in the lowest order can
then be written:

$$G_{\text{Umklapp}} = \sum_{p} \frac{1}{p} U(p)\, Q_i^p \qquad\qquad 11.17$$

where:

$$p\, \vec{k} = \vec{t^*} \qquad\qquad 11.18$$

Let us now return to our example of the most simple Landau-Ginzburg
energy in equation 11.3 and include the two Umklapp energies for Q_1 and Q_2:

$$G(Q_1,Q_2) = \frac{1}{2} A(T-T_c)\, Q_1 Q_2 + \frac{1}{2} g\, k^2\, Q_1 Q_2 + \frac{1}{4} B\, Q_1^2\, Q_2^2 + \dots$$

$$+ \Sigma\, \frac{1}{p} U(p)\, (Q_1{}^p + Q_2{}^p) \qquad\qquad 11.19$$

This Gibbs free energy can be written using equation 11.2 as a function of the amplitude Q_0 and the vector k:

$$G(Q_0,k) = \frac{1}{2} A (T-T_c{}^*) Q_0^2 + \frac{1}{4} B Q_0^4 + ...$$

$$+ \Sigma \frac{2}{p} U(p) Q_0^p \cos (p\phi) \qquad \qquad 11.20$$

The Gibbs free energy depends explicitly on the phase factor ϕ. The critical k-vector $k=k_m$ of the structural modulation follows then directly from the conservation law 11.18 and the stability of the lock-in phase is determined by the Umklapp energy $U(p)$. The stable solutions for the phase angle follow from the condition that $G (Q_0,\phi)$ has to be a minimum with respect to ϕ, leading to the solution:

$$\phi = 0, \frac{2\pi}{p}, \frac{4\pi}{p}, \frac{6\pi}{p}, ... \qquad \qquad 11.21$$

The amplitude of the order parameter Q_0 in a simple second order phase transition ignoring higher order terms in 11.20 follows from the minimisation of the Gibbs free energy with respect to Q_0:

$$A(T-T_c{}^*) Q_0 + B Q_0^3 + \Sigma \, 2U(p) Q_0^{p-1} \cos (p \phi) = 0 \qquad \qquad 11.22$$

For the typical case of p=4 (i.e. k=0.25) we find in the lock-in phase:

$$Q_0 = \sqrt{- A(T-T_c{}^*) / (B+2U(p) \cos \lfloor 4\phi \rfloor)} \qquad \qquad 11.23$$

The allowed values of cos [4 ϕ] are 1 and -1. The lock-in phase is more stable than the incommensurate phase if the Umklapp energy is negative (positive) and the phase angle ϕ= 0, $\pi/2$, π ($3\pi/2, \pi/4, 3\pi/4$).

The coupling between the modulation and the crystal lattice thus reduces the Gibbs free energy if a multiple of the wavevector of the modulation falls on a reciprocal lattice point thereby releasing Umklapp energy. This Umklapp energy, of order p, will then renormalise the energy of the p-th order term in the Landau potential. As we have already seen in the case of coupling between two order parameters in non-modulated materials, the effective order p of these coupling terms is not necessarily restricted to low values such as in the bare Landau potential and lock-in energies of the order six and higher seem to be rather common, although one might expect that the thermal stability range of lock-in phases with large values of p will be very small indeed.

Let us now return to the question of the stability of the structural modulation with respect to fluctuations. The lock-in mechanism with p>2 does not lead to any change of the longitudinal susceptibility (11.5, 11.6) and there is no change of the amplitudon behaviour, therefore. The phason dispersion, on the other hand, changes significantly. The dispersion curve in equation 11.8 is now related to the transverse frequency at the modulation wavevector k_m:

$$\chi^{-1} \propto \omega_T^2 \propto |U(p)| Q_0^{p-2} \qquad\qquad 11.24$$

which is finite and positive. This result demonstrates the physical nature of the lock-in phase: the lattice-modulation interaction locks the phase of the distortion wave so that the susceptibility of the phason is finite. The strength of the locking is proportional to the interaction energy.

11.2 Modulations in real space: the soliton lattice

In the preceding chapters we have analysed structural modulations in a thermodynamic sense without referring to specific atomistic features which might lead to the formation of modulated phases. The advantage of this approach is its generality, an essential advantage if we are interested in the thermodynamic description of complex materials such as the minerals or ceramics widely used in industrial applications. It is hard to believe that a fully satisfactory explanation about the exact structural meaning of the relevant order parameters will ever be given for these materials. The typical lack of complete structural information does not prevent us, however, from obtaining fundamental physical rules which lead us to a quantitative analysis of phase diagrams, sequences of phase transitions, the energetics of structural modulations, and the formation of macroscopic spontaneous strain in ferroelastic and co-elastic phases.

In co-elastic materials we can, however, be tentatively more specific about the underlying physical mechanisms which stabilise structural modulations without losing ourselves in detailed structural models. Experimental observations in these materials seem to indicate that one of the driving forces which leads to the formation of modulated phases is the self-compensation of the spontaneous strain on a mesoscopic scale (e.g. Harris et al. 1989,

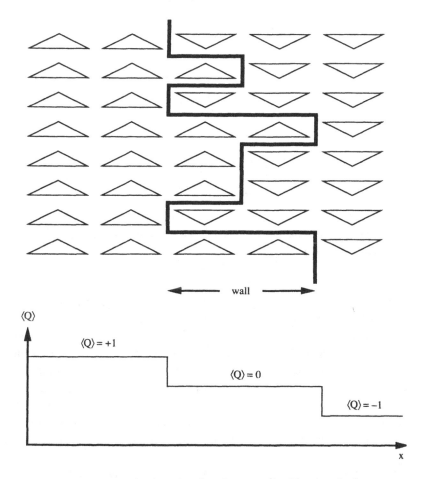

Fig. 11.8 Kink configuration in order-disorder systems with strong lattice distortions (e.g. $NaNO_3$, $CaCO_3$). The wall is characterised by rather random configurations with a finite width of the wall. In the lower half of the drawing the order parameter profile as integrated over planes parallel to the kink is depicted.

Carpenter 1988). The physical meaning of the order parameter Q_0 is related to an elastic lattice distortion in the same way as discussed in Chapter 5. The second ingredient for the formation of strain modulations is the gradient interaction between two order parameters as quantified by the Ginzburg energy in Chapter 11.1. The structural reason behind the negative gradient energy is that the structural configuration within the kink has a lower Gibbs free energy than the structure in regions away from the kink. In Fig.11.8 a kink configuration is depicted in which the homogeneous structure on the left

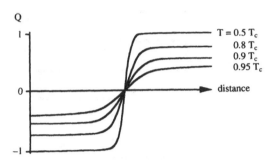

Fig.11.9 Profiles of domain walls in spin systems. Theoretical curves by Tentrup and Siems (1986) obtained from an ANNNI model.

and right hand side of the figure represents the undistorted co-elastic structure. This structure has, by definition, a lower Gibbs free energy than the paraelastic structure because we consider the co-elastic phase. The crystal structure inside the kink is, in terms of the elastic deformation, close to the structure of the high-symmetry phase and has thus an enlarged Gibbs free energy. The kink energy can be diminished and becomes even smaller than the energy of the homogeneous co-elastic material if this is helped by a further structural deformation, such as an instability of an optical mode, a cation ordering process or a defect ordering. The latter case is indicated in Fig.11.9.

We shall find this argument in Chapter 11.4 to be essential for the understanding of the Heine-McConnell model of incommensurations. At this point it is only important to understand that there are specific structural configurations which lead to negative gradient energies and thus to the tendency of the crystal to form as many kinks as possible. The total density of kinks is then limited by the repulsive kink-kink interaction. The physical nature of these interactions is far from being understood. It is commonly assumed that the interaction energy can be described as:

$$G_{kink\text{-}kink} = \kappa \exp(-r/R) \qquad\qquad 11.25$$

where R is the interaction length, r is the coordinate describing the intersoliton spacing and κ is the interaction strength (Fig.11.10).

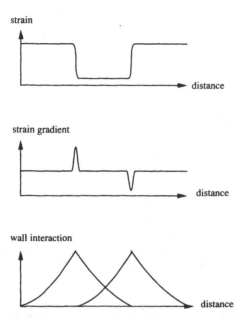

Fig.11.10 Profiles of periodic strain kinks, their gradient and the effective inter-kink interaction. Periodicity is often generated by the overlap between the potentials of the wall interaction which leads to a repulsive force between walls.

The Ginzburg energy is then equivalent to the negative wall energy which, for positive gradient energies, was derived in equations 8.39 and 8.52, and the positive inter-kink energy. The total energy becomes:

$$G_{Ginzburg} = L\{G_{Wall} + \kappa \exp(-r/R)\}/r \qquad\qquad 11.26$$

where L is the total length of the crystal. This function is plotted in Fig.11.11. The equilibrium wavelength of the soliton lattice is large compared with the unit cell dimensions of the underlying lattice. We would thus expect that strain induced modulations occur at rather small k-vectors of the acoustic phonon branch, similar to the behaviour of quartz, cordierite or kinetically disordered feldspar.

So far we have seen that strain modulations lead to periodic structures with periodicities measured in several tens of Ångstroms, which are superimposed on the underlying structural lattice with periodicities of some Ångstroms. The essential feature which allowed the gradient energy to become negative was the capacity of some crystal structures to provide an energy lowering

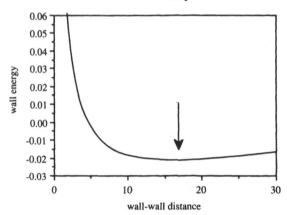

Fig.11.11 Effective interaction potentials (equation 11.2.2). The wall energy is given in arbitrary units. The equilibrium point is marked by an arrow. The wall distance is measured in numbers of unit cells.

mechanism for the domain wall containing structural elements which do not exist in the co-elastic material (Fig.11.9). In order to take full advantage of this mechanism, a large proportion of the kinks must be located on the appropriate positions in the crystal structure. All other kinks will not participate in the lowering of the Ginzburg energy because they still possess positive gradient energies. The condition of optimal overlap between the strain modulation and the periodicity of the crystal structure thus directly generates lock-in phases in which such an overlap exists. The stabilisation energy of the lock-in phases is then exactly identical to the Umklapp energy described in equation 11.17 and we expect that most strain modulations are, indeed, commensurate modulations rather than incommensurations.

There is, however, competition between the periodicity of the soliton lattice derived from the minimization of their internal energies in equation 11.26 and the commensuration determined by the Umklapp energy in equation 11.17. If these two periodicities do not match, we expect a third level of periodicities due to proper incommensurations. These periodicities, on an increasing length scale, can then be classified as:

1 The periodicity of the crystal lattice as the smallest length scale;
2 The strain modulation leading to the soliton lattice on the next larger length scale and;
3 Incommensurations of the soliton lattice which, as we shall see, form another soliton lattice on an even larger length scale.

These latter incommensurations will occur only if the inter-kink forces are strong enough to compete with the Umklapp energy. If, on the other hand, the inter-kink energy is small we expect local commensurations to break the periodicity of the soliton lattice in class 2 which can ultimately lead to chaotic states as often observed in kinetically generated modulations (see Chapter 12).

Let us now consider the strain incommensurations referred to in point **3** above in the limit of large inter-kink energies. We can write the total Gibbs free energy for a Landau potential with a quadratic and a positive fourth-order term as in equation 11.1:

$$G = \Sigma \left(\frac{1}{2} A(T-T_c) + \frac{1}{2} g k^2\right) Q_1 Q_2 + \frac{1}{4} B Q_1^2 Q_2^2 + \Sigma \frac{1}{p} U(p) \left(Q_1^p Q_2^p\right) \qquad 11.27$$

The relevant wavevector for the soliton lattice in class 3 is the deviation of k from the lock-in wavevector $k_0 = \tau/p$ where τ is a vector of the reciprocal lattice of the underlying crystal structure. This difference vector is Δk. The order parameter is again complex and can be written in terms of the amplitudon and phason coordinates (equation 11.10 and 11.11). As the amplitude of the modulation is essentially determined by the local strain it will not change much for the incommensurations and we can concentrate on the phason part of the Gibbs free energy alone (i.e. constant amplitude approximation). The relevant part of the Gibbs free energy is then:

$$G_{phason} = - A^2/B \ (T_c-T)^2 + G_{soliton} \qquad 11.28$$

with

$$G_{soliton} = \frac{1}{2} g \, Q_0^2/v \int G(r) \, dr \qquad 11.29$$

where v is the total volume, Q_0 is the amplitude of the order parameter and $G(r)$ is the space dependent part of the free energy including the Umklapp energy U' :

$$G(r) = \frac{1}{2} \left(\nabla \phi \, (r) - \Delta k\right)^2 - U' \cos \left(p \, \phi \, (r)\right) \qquad 11.30$$

The Umklapp-energy U' accounts for the prefactor of $G_{soliton}$ in equation 11.29, namely:

$$U' = U \ Q_0^{p-2}/g \qquad\qquad\qquad\qquad\qquad\qquad 11.31$$

We can now solve the integral in equation 11.29 which determines the soliton energy in one dimension and find:

$$G_{soliton} \propto \ 1/L \int \left\{ \frac{1}{2} \left(d \ \phi/dr\right)^2 - U' \ (\cos p \ \phi \ (r) - 1) \right\} dr$$

$$- \Delta k/L \ (\phi \ (L) - \phi(O)) + \frac{1}{2} \Delta k^2 - U' \qquad\qquad 11.32$$

The nature of the incommensurate phase is now determined by the function $\phi(r)$ which minimizes the soliton energy 11.32. The only part of this energy which depends explicitly on ϕ r) is the integral in 11.32 and Bak and Emery (1976) have shown that the condition for an extremum of this integral:

$$d^2 \ \phi(r)/dr^2 = p \ U' \sin p \ \phi(r) \qquad\qquad\qquad 11.33$$

is the well-known Sine-Gordon equation of solitons with its solution for each soliton:

$$\phi(r) = 4/p \ \tan^{-1} (\exp \left[p \ \sqrt{U'} \ r \right]) \qquad\qquad\qquad 11.34$$

As each phase change due to a soliton adds to the previous phase changes, we can write the total phase change at a position r in the crystal after m solitons have passed as:

$$\phi_{total}(r) = \phi_0 + 2\pi \ m/p + 4/p \ \tan^{-1} (\exp (p \ \sqrt{U'} \ r)) \qquad 11.35$$

This phase function is shown in Fig.11.12 (right hand side). It describes periodic phase changes between the lock-in values of multiples of $2\pi/p$. Commensurate soliton lattices (i.e. strain modulations in class 2) are represented by the plateaus. The corresponding modulation in real space is also depicted in Fig.11.12.

We have assumed so far that the amplitude function Q_0 does not vary in space. This assumption was necessary in order to treat the mathematical problem of minimizing the total Gibbs free energy in a simple way. Blinc et al. (1983) have shown however that small amplitude variations do exist which basically follow the second derivative of the phase angle:

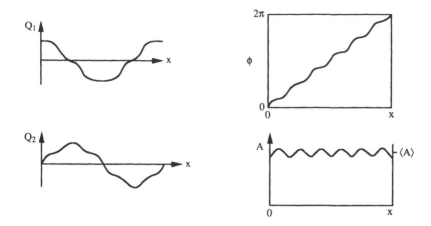

Fig.11.12 Spatial variation of Q_1 and Q_2 as strongly coupled order parameters in equation 11.2.3. Both order parameters can be expressed by the amplitude and phase function (right hand side) as $Q_1 = A \cos \phi$ and $Q_2 = A \sin \phi$.

$$\frac{\partial Q_o}{\partial r} \propto \frac{\partial^2 \phi}{\partial r^2}$$

11.36

In Fig.11.12 the phase angle, the amplitude modulation and the two basic order parameters Q_1 and Q_2 are shown as calculated numerically from the equation 11.27 by Blinc et al (1983).

We finally return to the soliton density in class 3. This soliton density results from the minimization of the Gibbs free energy with respect to the inter-soliton spacing in the same way as the treatment of solitons in class 2. Unfortunately, no simple analytical solution exists beside the obvious fact that the soliton density increases with increasing temperature and is a maximum at the transition point between the high-symmetry phase and the modulated phase. This is also the point at which the wall thickness is largest and it is of the same order of magnitude as the inter-soliton spacing. At lower temperatures, the inter-soliton spacing increases and becomes large compared with the decreasing wall thickness near the transition point to the homogeneous low-symmetry phase. In the absence of defects, we expect then the soliton lattice with class 3 periodicity to decompose into randomly spaced solitons which are individually pinned by the crystal lattice. This means that the long range order of the soliton lattice is destroyed and chaotic states with a high degree of short range order might still persist in the low-symmetry

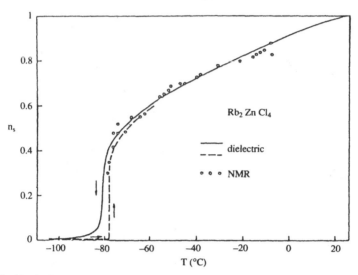

Fig.11.13 Temperature dependence of the soliton density in Rb$_2$ZnCl$_4$ as derived from dielectric and NMR data. The finite value of n$_s$ at T$_c$ and in the C phase below T$_C$ demonstrates the presence of the metastable chaotic state. (After Blinc et al. 1983).

phase at T< T$_c$. The experimentally observed soliton density normalized to unity at the transition point between the high-symmetry phase and the modulated phase is shown in Fig.11.13.

11.3 Return to the Landau-Ginzburg energy in reciprocal space and the Heine-McConnell model

We have seen from the treatment of strain modulations in direct space that the basic physical features of the modulation are described by soliton lattices with wavelengths which are, in general, large compared with the lattice parameters of the underlying crystal structure. We have also seen that an essential ingredient of the modulation is the negative gradient energy as part of the Ginzburg energy (11.1). The third finding was that the two order parameters involved, namely Q$_1$ and Q$_2$, have phase angles which are shifted by π/2, independent of their actual waveform. Let us now concentrate on the description of the collective character of the soliton lattice rather than its structural properties in real space and follow Heine and McConnell (1984) in their treatment of order parameter coupling in a two-dimensional order parameter vector space including gradient coupling.

We may assume that the physical meaning of the two order parameters Q_1 and Q_2 can be very different indeed, such as an elastic shear for Q_1 and a cation ordering process for Q_2. What is essential is simply that the gradient energy when rescaled in one order parameter has to be negative. This can best be illustrated if we go back to the coupling between two order parameters in Chapter 9 where each order parameter is a complex quantity. We then include the gradient coupling terms introduced in Chapter 11 and write a Gibbs free energy of two complex order parameters Q_1 and Q_2 with the usual condition of the k-vector conservation $\Sigma\, k_i = 0$ as:

$$G(Q_1Q_2) = L(Q_1) + L(Q_2) + G_{Ginzburg}(Q_1) + G_{Ginzburg}(Q_2)$$

$$+ \zeta_1\,(Q_1Q_2{}^* + Q^*{}_1Q_2) + \zeta_2\,(Q_1{}^*Q_1Q_2{}^*Q_2) + ...$$

$$+ \zeta_{gradient}\,(Q_1\,\nabla Q_2 - Q_2\,\nabla Q_1) \qquad\qquad 11.37$$

where L is the Landau potential containing only the real quantities QQ^*, $(QQ^*)^2$ etc. and real coefficients. The Ginzburg energy is given in equation 11.1 and the last three terms represent the coupling energies as discussed in Chapter 8. The last term is the Lifshitz energy which describes the gradient interaction of the two order parameters as introduced in Chapter 11. We can now follow our previous description of the modulated structure in which one order parameter followed the gradient of the other order parameter, namely:

$$Q_1 \propto \nabla Q_2 \text{ and } Q_2 \propto -\nabla Q_1 \qquad\qquad 11.38$$

with the same proportionality constant for both equations.

The gradient coupling can then be rewritten as the Ginzburg energy:

$$\zeta_{gradient}\,(Q_1\nabla Q_2 - Q_2\nabla Q_1) \propto \zeta_{gradient}\,\{(\nabla Q_1)^2 + (\nabla Q_2)^2\} \qquad 11.39$$

A negative gradient coupling renormalizes the Ginzburg energy to smaller values and can lead ultimately to a negative total gradient energy. The uniform bi-linear coupling will then lead to small phase changes and we can rewrite the total Gibbs free energy ignoring higher order coupling as a function of only one complex order parameter (say Q_1) and a negative Ginzburg term, identical to the starting equation for the soliton lattice in equation 11.4. We have thus found that the gradient coupling between Q_1 and Q_2 in the Heine-McConnell model leads, in fact, to the same reduction of wall energy as in the soliton-lattice picture in Chapter 11.1.

Let us now describe the dispersion relationships of the two order parameters Q_1 and Q_2 in reciprocal space. A Fourier transform of the Gibbs energy in equation 11.37 ignoring higher order coupling leads to:

$$G = 1/2\, A_1\, (T-T_{c1})\, Q_{1o}{}^2 + 1/4\, B_1\, Q_{1o}{}^4 + \ldots$$

$$+\ 1/2\, A_2\, (T-T_{c2})\, Q_{2o}{}^2 + 1/4\, B_2\, Q_{2o}{}^4 + \ldots$$

$$+\ 1/2\, g_1 k^2\, Q_{1o}{}^2 + 1/4\, g'_1\, k^4\, Q_{1o}{}^4 + \ldots$$

$$+\ 1/2\, g_2\, k^2\, Q_{2o}{}^2 + 1/4\, g'_2\, k^4\, Q_{2o}{}^4 + \ldots$$

$$+\ \lambda_1\, Q_{1o} Q_{2o} - 2k\, \zeta_{gradient}\, Q_{1o} Q_{2o} + \ldots \qquad\qquad 11.40$$

where Q_{1o} and Q_{2o} are again the amplitude functions. The first four terms are the usual Landau energy, the four following terms are the bare Ginzburg energies for the two order parameters and the last two terms are the uniform and the gradient coupling energies, respectively. The expresssion 11.40 is identical to the Gibbs free energy of two order parameters with bi-linear coupling between them as discussed in detail in Chapter 10.3. The only difference is now that the coupling parameters depend explicitly on the wavevector k. The general solution for the thermodynamic equilibrium condition applied to this Gibbs free energy is thus identical to the solution in equation 10.19. This solution is greatly simplified if we consider a physical situation in which the two critical temperatures are different so that only one of them leads to a structural phase transition between a ferroelastic or co-elastic phase and a strain-modulated phase. The second critical temperature is then considered to be much higher and can be ignored for this discussion. This is the essential reason why Heine and McConnell could distinguish between the 'main' soft-going mode, say Q_1 related to the co-elastic strain, and the 'subsidiary' mode, say Q_2. Note, however, that this distinction becomes impossible when the two critical temperatures approach each other so that the temperature evolution of $A_2(T-T_{c2})$ becomes noticeable.

Ignoring thus the bare entropy of the subsidiary order parameter, Q_2, we can describe the coupling in reciprocal space. Let us assume that Q_1 is closely related to the spontaneous strain. The bare dispersion of Q_1 is then essentially that of an acoustic phonon. The dispersion of Q_2 is non-critical and can be assumed without loss of generality to have its minimum at the origin of the Brillouin zone. The two bare modes are shown in Fig.11.14 as dotted lines.

The bilinear coupling leads to an overall energy reduction and depends linearly on the k-vector thus reducing the acoustic phonon branch at k-vectors

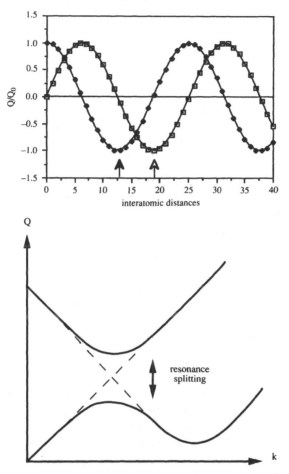

Fig.11.14a) Two order parameters Q_1 and Q_2 shown as harmonic waves with strong gradient coupling. The two waves are phase shifted so that the extremum of one wave coincides with the zero point value of the other wave. The strength of the coupling depends on the gradient of the order parameter and, thus, on the wavevector of the modulation (equation 11.3.7).
b) Equivalent splitting of the energy dispersion branches due to the coupling between the two order parameters. This coupling effect is called 'resonance splitting' by Heine and McConnell (1984).

inside the Brillouin zone. The temperature evolution of the effective dispersion of the Q_1-like branch (Fig.11.14b) stems from the T-dependence of the bare Landau potential in Q_1. This Landau potential would lead to a structural phase transition between a paraelastic phase and a co-elastic phase at a temperature T_c. As discussed in Chapter 5, this phase transition is

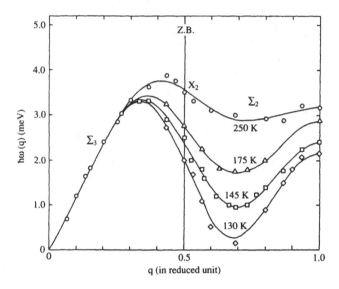

Fig.11.15 Softening of the acoustic phonon branch near the critical wavevector k=0.69 mm. (After Iizumi et al. 1977).

correlated with an elastic anomaly with a softening of the acoustic phonon branch. This softening means that the slope of the Q_1 dispersion at the origin of the Brillouin zone decreases and thus drives the negative hump towards lower energies. A structural phase transition between the paraelastic and a modulated phase occurs when the minimum of the Q_1 dispersion reaches the zero-energy level. The modulated phase will transform to the co-elastic phase when, at even lower temperatures, the slope of the Q_1 dispersion at k=0 disappears. It is obvious that the same symmetry restrictions as discussed before for bilinear coupling and the k-conservation law applies for the Heine-McConnell model and the reader is referred to their review article for further discussion.

Finally, I would like to stress the point that the description of strain modulations in reciprocal space is a convenient way to display thermodynamic properties rather than real phonon dispersions. Only if the order parameters in the Gibbs free energy in equation 11.40. represent amplitudes of quasiharmonic oscillations, such as in purely displacive phase transitions, can we associate the dispersion curves in Fig.11.14 with acoustic and optic phonon branches. In most cases, however, no such oscillatory motions exist and the dynamical behaviour of the order parameter response is diffusive or even related to transport phenomena rather than to local vibrations. This makes a dynamical dispersion relation unphysical. The dispersion relations used in this chapter, however, have nothing to do with the dynamical response of the order parameter but they simply represent the wavevector-dependence of the Gibbs free energy related to bare and renormalized order parameters. Although both pictures must be identical for two phonon driven order parameters, the similarity with other physical situations is almost fortuitous.

Let us now consider the behaviour of paraelastic crystals at $T \geq T_C$. Many paraelastic crystals show the following features under thermodynamic conditions close to the transition point:

1 The phonon dispersion of the acoustic branch shows a negative dip as described in the last chapter even if the actual low-symmetry phase is not modulated (Fig.11.15).

2 The microstructure may consist of strain related tweed patterns as shown in Fig.11.16. Although we shall see that these tweed patterns are a characteristic feature of non-equilibrium behaviour (Chapter 12) it appears that the lifetime of these patterns is exceedingly long in ferroelastic and co-elastic materials so justifying their description as precursor order.

3 The diffraction pattern of such precursor material shows diffuse diffraction signals which could be indexed with respect to the low-symmetry phase. The matrix of the crystal is, however, unstrained so that the crystal is still paraelastic, i.e. there is no spontaneous strain present. The positions of the superlattice reflections are not necessarily compatible with the lattice of either phase and they can thus not be indexed with respect to the uniform stable states (Fig.11.17, see Salje et al. 1987 and references given therein).

Although much attention has been given to these phenomena in martensitic alloys, little research has been undertaken in non-metallic materials. The strain-spinodal treatment by Suzuki and Wuttig (1975) showed the possibility of the existence of microstructures with low symmetry structure in the paraelastic phase around lattice imperfections.

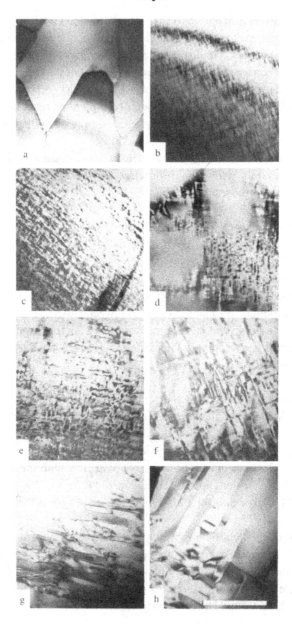

Fig.11.16 Microstructures in Cordierite as observed in a kinetic experiment. The
homogeneous crystals in (a) are the hexagonal material without strain modulation.
This phase transforms in a first-order transition kinetically into modulated crystals
(b,c,d) which show the typical tweed pattern. Orthorhombic cordierite (e,f,g,h) is
highly twinned with microstructures rather similar to that of the modulated state.
The macroscopic strain builds up only in the orthorhombic phase. At 1400°C the

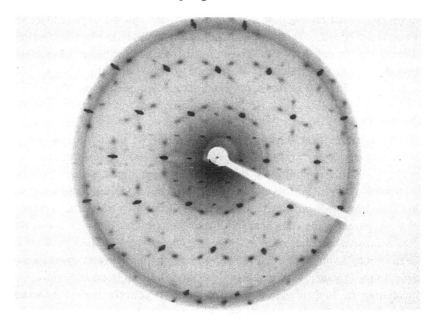

Fig.11.17 Precession photograph of $Pb_3(PO_4)_2$ doped with 8% Ba replacing Pb. The strong reflections relate to the paraelastic phase $R\bar{3}m$. The additional weak reflections are streaked and show diffuse reflection lines forming hexagonal stars around the main reflections. The additional reflections relate to monoclinic microdomains in the trigonal matrix. Although each microdomain generates strain, the macroscopic strain on a scale larger than 100Å is zero.

Molecular dynamic calculations by Clapp et al. (1988) arrived at tweed-like microstructures even without the explicit evocation of an underlying defect structure. Yamada et al. (1985) introduced the name 'embryo' for microdomains of low-symmetry material in a high-symmetry matrix. The occurrence of such embryos defines the precursor regime. As their treatment is closest to our own line of argument we shall use here their nomenclature as much as appropriate.

Fig.11.16 (cont'd) transformation is a→b→c→d. At lower temperatures (1280°C,1190°C) the sequence is a→b→c→e→f→g→h avoiding the coarse nucleation in d. (After Putnis et al. 1987). (scale bar = 0.2μm).

All the structural phenomena reported so far can be explained if we consider the crystal in the paraelastic phase to contain long-lived fluctuations of the order parameter. We have seen in Chapter 11 that these non-uniform fluctuations in the paraphase contribute to the Ginzburg energy. The magnitude of the fluctuations depends on the gradient term g (equations 11.1, 9.22 and 9.23) and on the susceptibility of the order parameter (equations 11.5 and 11.6).

The basic question then is how the lifetime of these fluctuations is increased with respect to the free fluctuations considered so far. One common mechanism is the coupling of the structural fluctuations with defects, as analysed in Chapter 9.3. As the lifetime of defects is virtually unlimited we find that defect induced fluctuations will also exist on a time scale which is long compared to the time scale of elastic relaxations. Fluctuations of the order parameter will thus couple with the strain field on a mesoscopic scale in the same way as coupling occurred before with the macroscopic spontaneous strain. In the language of Yamada, we call strain-related fluctuations 'dressed' fluctuations (a fluctuation is called **dressed** if it is accompanied by strain fields). These fluctuations have lower Gibbs free energies (i.e. reduced by the elastic energy and the coupling energy) than the bare fluctuations of the order parameter which would occur if the elastic deformation could not follow rapid fluctuations.

Let us now consider the local correlations of the precursor fluctuations in real space. The three following mechanisms seem to dominate the short range order in the paraelastic phase.

1 Random fluctuations with a predominantly quadratic Ginzburg energy. The long-wavelength fluctuations are thus more probable than short wavelength fluctuations because of the k^2-dependence of the Ginzburg energy in equations 11.1 and 9.23. A white noise spectrum will occur only if the gradient term g disappears: in all realistic cases we find a rather substantial k-dependence of the noise spectrum.

2 Dressed fluctuations must be anisotropic according to the 'soft directions' of the elastic constants. This leads to a decrease of randomness of the shape of the fluctuating clusters. If these clusters show a low degree of correlation with each other, we call them thermally induced 'embryos'. At temperatures well above the transition point, bare embryos are frequently created but are soon anihilated because their sizes do not exceed the critical radius as defined in Chapter 8.1. These embryos are bare because their lifetime is too short for lattice relaxation to follow. At temperatures closer to the transition point, the lifetime increases and dressing occurs. The relevant lattice relaxations

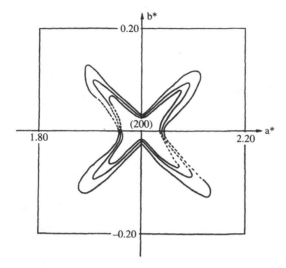

Fig.11.18 Huang scattering in Fe-Pd alloy. The scattering intensity in the arms of the star is related to embryonic fluctuations. The intensity was obtained by Seto et al. (1989) by substraction of the data at 315K (T>>T_c) from the data at 275K, just above T_c. (see Fig.8.23).

stabilise the embryos, each of which is now described by order parameter and strain solitons as in Chapter 8. For low embryo densities their mutual interaction is weak and we expect their random distribution in the paraphase. A typical example of complex diffraction patterns including anomalous Huang scattering is shown in Fig.11.18 for bcc-based alloys (Seto et al. 1989).

3 Large fluctuation densities will lead to an overlap of soliton walls and strong interactions between the dressed embryos. As the elastic lattice distortion is long-reaching, self-organisation will occur even over large interatomic distances. The most common microstructure is the strain modulation as shown in Chapter 8.4 where periodic modulations with self-compensating strain patterns are surrounded by one common strain field. As shown in equation 8.64, the wavevectors of these modulations point along the mechanically soft directions in the structure and they are thus normally orthogonal. The wavelength of the modulation depends on the strength of the intersoliton interaction and shows more randomness than the orientation of the wavevector of the modulation.

This microstructure changes during the structural phase transition. These changes are, however, less concerned with the individual solitary waves than with their mutual interaction. We have seen in Chapter 8 that random fluctuations and modulations are described by the same kink solitary waves and 'soliton lattices'. In the low-symmetry phases, these kinks, when static, become twin walls with wall thickness decreasing with decreasing temperature. In a second order phase transition, the changes from the precursor regime to the low-symmetry phase are rather continuous, leading from weak fluctuations via modulations, to twinning (Fig.11.16, Fig.7.5). Tweed patterns typically occur in the high-symmetry phase because periodic kinks with orthogonal wavevectors interact only weakly. In the low symmetry phase, on the other hand, such interactions are stronger and parallel twin patterns with few intersections are characteristic.

The same two orientations of the kink-walls persist in the low symmetry phase as the dominant twin planes and the twin planes are thus parallel to the soft kinks in the precursor regime. The twin walls in the low symmetry phase strongly interact leading to the microstructures described in Chapter 7. This inter-kink interaction increases the Gibbs free energy and the superposition of the two kink-lattices is energetically unfavourable. As a consequence, we find that the two kink-lattices are subject to a microstructural avoidance rule: competing kink-lattices seperate from each other and tend to minimize their degree of interconnection. The typical microstructure in the low-symmetry phase close to the transition point consists of patches of kink-lattices with little mutual penetration and strain related microstructures at their interconnection (see Chapter 7). In Fig.7.5 these drastic changes of the microstructures due to the structural phase transition are shown with intersecting kink-lattices (tweed) in the high-symmetry phase, patches of non-interpenetrating kink-lattices with large structural distortions on their joints at temperatures near the transition point, and coarse twin patterns in the low-symmetry phase.

Let us now recall the essential structural features which allow kinks to penetrate in the high-symmetry phase but repel each other in the low symmetry phase. We return to the discussion of the strain-solitary waves in the case of a ferroelastic phase transition. We have found that the structural state in the middle of the node of a kink is identical with the uniform high-symmetry form. Steps in the transition mechanism (Chapter 8.2) will extend this part of the kink and lead to double kinks joining the order parameters $+Q$ and $-Q$ through a finite interval with $Q=0$. The local switching process at $+Q$ or $-Q$ with a local tetragonal structure, for example, in Chapter 8.2, was shown to be of orthorhombic symmetry. The intersection of the second kink now performs this switching: one tetragonal form in the first kink is

transformed into the second tetragonal orientation in the second kink via the orthorhombic strain. As the tetragonal kinks are periodic (although with large fluctuations of the periodicity) the intersections show the same periodicity and it is a question of taste whether we prefer to describe the observed microstructure as an array of tetragonal kinks with orthorhombic intersections or as an array of orthorhombic knots with tetragonal strings between them.

The model in Fig.11.19 shows a tweed microstructure, with the tetragonal and orthorhombic parts indicated, as built up from solitary waves as shown in Chapter 8 (e.g. 8.41, 8.45 and 8.50).

Both parts possess identical energy because the order parameter of this phase transition is double degenerate with the two basis functions representing the tetragonal and the orthorhombic spontaneous strain. Kink interactions are thus energetically permissible and represent a structural modification which has the same energy as the kink itself. The degeneracy between the two structural states is lifted in the low-symmetry phase and the kink-energy is lower than the intersection energy. The kinks thus repel each other via the elastic interaction energy discussed in Chapter 7.

The same argument has been used by McConnell (1971) for the analysis of the microstructure of modulated adularia (Fig.8.27). Here the two degenerate structural states are compression in the kinks and shear in their intersections. McConnell has also extended this model for the interpretation of precursor modulations in alloys and it appears that this mechanism of structural degeneracy in the fluctuation regime is one of the most elegant ways to explain the observed long life-time of modulated fluctuations. Extending the same argument to the discussion of higher order Ginzburg energies (Chapter 11) we see that terms of the fourth order represent a biquadratic coupling between the two fluctuating order parameters:

$$G_{Ginzburg} (Q_1, Q_2) \propto (Q_1)^2 (Q_2)^2 k^4 \qquad 11.41$$

with a coupling constant which depends explicitly on the wavevector k. A negative coupling energy leads then to a decrease of the phonon dispersion of the acoustic phonon branch as discussed above.

We finally describe long-lived modulation in the precursor regime in the order parameter vector space, as introduced in Chapter 10. As in the discussion of the cubic-to-tetragonal phase transition in a ferroelastic or co-elastic crystal, we analyse the surface of the free energy at $T > T_c$ with a global

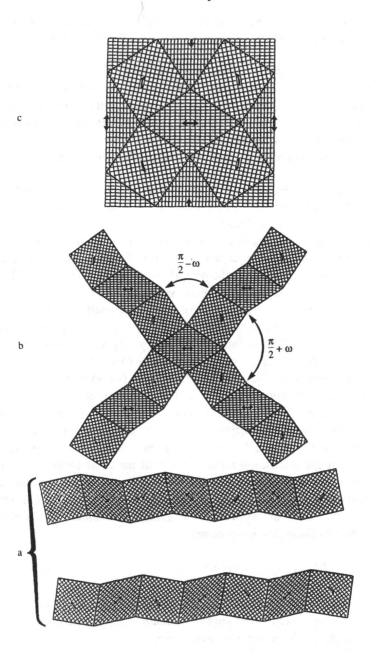

Fig.11.19 Structural elements leading to the formaton of tweed patterns. Kink lattices consisting of rotated elements with Q=0 alternate with Q=+Q_0 and Q=-Q_0 in (a). As the two distortions +Q_0 and -Q_0 are not equivalent, the deformation pattern in the two ribbons in (a) are also not equivalent. Taking one diamond shaped

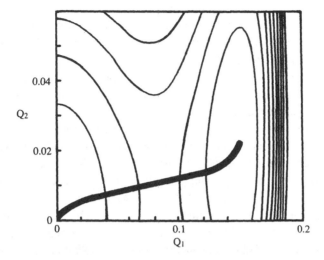

Fig.11.20 Potential surface for a first order phase transition at $T=T_c$. A 2-4-6 Landau potential governs Q_1, a purely parabolic potential was approximated for Q_2 (e.g. representing the lattice strain). The coupling is linear-quadratic $\propto Q_2Q_1^2$. The bold line represents the trajectory between the two stable solutions at its end points. Note the saddle point in the middle of the trajectory (after Fuchizaki and Yamada 1989).

minimum at $Q_1=Q_2=0$. Local side minima occur either due to the fact that the phase transition is first order, or may result from dips in the dispersion curve or any other microscopic phenomenon. In the example of Fig.8.12 in Chapter 8.2 we assume that this tie-line is linear under the given thermodynamic conditions (see Barsch and Krumhansl 1984). This linearity will, in general, not apply and we shall find curved trajectories as indicated in Fig.11.20. In either case, the linear or non-linear trajectory in order parameter vector space leads to a strong coupling between the two order parameters and thus to kink profiles of the same type as discussed before for $T<T_c$.

Fig.11.19 (cont'd) element of both ribbons as a common element, both ribbons intersect with orthogonal 'wavevectors'. The structural elements linked to the intersection do not form a star with orthogonal arms but possess angles between them of $\pi/2 \pm \omega$ (see Chapter 8). The stars in (b) can now be superimposed for all shear areas ($\pm Q$) forming the tweed pattern in (c). The final tweed can then be understood as interpenetrating ribbons or as a tiling of rotated (e.g. tetragonal) and sheared (e.g. orthorhombic) structural elements.

In Fig.11.21 the results of numerical calculations by Seto et al. (1989), for the case of Fe-Pd alloys are shown. This example illustrates the splitting of one kink between two states related to the low-symmetry phase, into two kinks with an intermediate plateau. This intermediate plateau represents any metastable state which is represented by a point on the trajectory in Fig.11.21. This metastable state does not necessarily appear in a macroscopic thermodynamic phase diagram because it might be stabilised by the fact that the slab of material in the middle of the kink is surrounded by parts of the crystal with a more stable structural configuration. It is thus the interaction between different structural states of the crystal, which occur simultaneously in a kink-solitary wave which stabilises the metastable state. The stabilising rate of the kinks can be enhanced if several kinks form a soliton lattice as described in Chapter 11.2. We can then represent these states in the energy diagram in Fig.11.22. The ordinate in this plot is the trajectory in Fig.11.21 and the absissa is the Gibbs free energy of the corresponding structural state. We see that the intermediate states are highly metastable and that the more stable configurations are in fact not reached if the interaction represented by the spring is strong enough. Barsch and Krumhansl (1988) and Falk (1983) have analysed this structural situation more closely and the term of a 'crest-riding periodon' is commonly employed for its description. How far the crest-

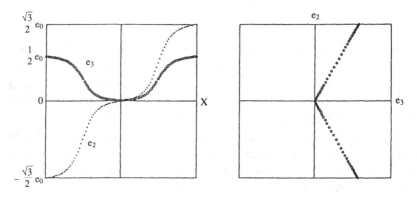

Fig.11.21 Strain profiles of the components e_2 and e_3 of the spontaneous strain (left) and the equivalent order parameter vector space (right). The stability points are $e_2=0, e_3=0$ and $\left(e_2 = \pm \frac{1}{2} \sqrt{3} \, e_0, \ e_3 = \frac{1}{2} e_0\right)$. The profile is plotted for $T \sim T_C$ with:

$$I(e_2, e_3) = \frac{1}{2} = A(T-T_c)\left(e_2^2 + e_3^2\right) + \frac{1}{3} B\left(e_3^2 - 3 \, e_2^2 \, e_3\right) + \frac{1}{4} C\left(e_2^2 + e_3^2\right)^2 \text{ and}$$

$$G_{Ginzburg} = \frac{1}{2} g \left\{\left(\nabla e_2\right)^2 + \left(\nabla e_3\right)^2\right\} \text{(after Seto et al. 1989).}$$

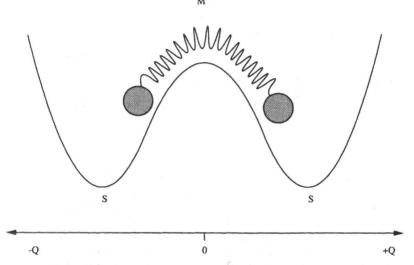

Fig.11.22 Simple model for the crest riding periodon. The crystal is cooled down from T>T_c with the local maximum at M as the stable configuration. At T<T_c, each minimum s becomes the stable state. If both states are sufficiently coupled, periodons can develop which contain interacting local states which do not themselves represent thermodynamics equilibrium but still lower the energy of the unstable Q=0 phase to a possibly metastable periodon 'phase'.

riding periodon represents a metastable or unstable structural configuration still seems to be a matter for debate, but even if the final answer is in favour of the unstable solutions, it still remains a structural state with a long life-time and it is thus a relevant feature of the short range order.

12

SOME ASPECTS OF THE KINETIC BEHAVIOUR OF FERROELASTIC AND CO-ELASTIC CRYSTALS: AN OUTLOOK

Structural relaxations can be slow if thermally activated processes such as cation ordering are coupled with the structural order parameter. If the wall energies are small, nucleation processes will dominate the kinetics of the structural relaxation. In ferroelastic and co-elastic crystals one often observes that such wall energies are high and a more continuous kinetic behaviour occurs in these materials. Some aspects of the general theory are outlined and examples given for the case of Al,Si ordering, coupled with a displacive distortion of a framework structure.

Non-equilibrium states in crystals are often the result of rapid changes in the thermodynamic conditions of the crystal, such as shock-heating or quenching, sudden changes of pressure, the rapid application of electric fields etc. We shall now consider the behaviour of crystals which tend to relax into the equilibrium state. If the lattice relaxation is completely governed by phonon processes, the relaxation time will be of the order of the phonon time itself, i.e. some 10^{-12} sec, which is too fast to be examined using classical experimental techniques such as X-ray diffraction, Mossbauer experiments, NMR etc. Slower relaxations are, however, common in nature with order parameters related to cation ordering, impurities, molecular disordering or other thermally activated processes. In order to illustrate this behaviour let us return to the example of the phase transition C2/m-C$\bar{1}$ in Na-feldspar (Chapter 10). The displacive order parameter, Q, describes a ferroelastic transition which interacts with the Al,Si ordering, Q_{od}, via bilinear coupling. Although the displacive process leads to relaxations on the phonon timescale, the total thermodynamic equilibrium state is only reached after relaxation of the slower order parameter Q_{od}. Below about 700K, the relaxation of Q_{od} is exceedingly slow and equilibrium might not be reached on a laboratory time scale at all. Similar observations were reported for the transition in LiKSO$_4$ (Balagurov et al. 1986) and it is also commonly observed that defect relaxations can be sluggish and may lead to kinetic hindrance during ferroelastic phase transitions. In the case of structural modulations, as in incommensurate phases, the strength of the coupling between the 'fast' order parameter and the defects need not be as strong as in homogeneous phases because even weak defects will act as phason traps with an effective pinning volume given by the Onstein-Zernicke potential (9.28). It is interesting, and necessary for the understanding of such non-equilibrium features, to explore the relevance of kinetic behaviour of ferroelastic and co-elastic crystals when coupling with slow processes occurs. For a review of a multitude of experimental observations in this field the reader is referred to Carpenter and Salje (1989).

Some general features of the relevant rate laws will now be discussed before describing specific results. Let us start with the following situation: a crystal has been shock-heated so that its structural state no longer represents a thermodynamical equilibrium state. The crystal will then tend to lower its Gibbs free energy by undergoing structural changes. These changes define the 'kinetic pathway' and it will be one of the tasks of the kinetic approach to identify this pathway. Each step on this pathway towards equilibrium is correlated with changes of a multitude of structural parameters (almost all atomic positions may be involved in the relaxation) and we must first identify the essential parameters which allow us to find the most probable pathway the system will take. Glauber (1963), Kawasaki (1966), Salje (1988) and others have argued that, by a judicious selection of parameters, kinetic expressions which describe the time evolution of a **kinetic order parameter** can be obtained. This kinetic order parameter is then the amplitude function of the kinetic transformation pattern in the same way as the equilibrium order parameter is the amplitude function of the transformation pattern in thermodynamic equilibrium. Experimental observations so far seem to indicate that the kinetic order parameter is often related to the same transformation pattern as the equilibrium order parameter (Carpenter and Salje 1989). In these cases, a clear distinction between both types of order parameter is not necessary and we shall use the term 'order parameter' as before for both the equilibrium and the kinetic variety. Ths second conjecture is that the kinetic pathway charts the passage of the system over a well defined Gibbs free energy surface towards the minimum point. This passage is the better defined the larger the penalty the system has to pay if it deviates from the pathway, i.e. the higher the energy barriers on either side of the pathway. This means that the pathway is topologically speaking, a valley in energy space which has its bottom at the equilibrium point (Fig.12.1). Such valleys have a close relationship with those of the order parameter vector diagrams as discussed in Chapter 10. The major difference is, however, that the 'fast' degrees of freedom, e.g. the strain, optical phonons etc., relax rapidly and that the time limiting steps are those of the slow order parameter under partial equilibrium of the fast order parameter.

Although this picture of a well-defined pathway is clearly incorrect for systems with kinetic bifurcations, chaotic states etc., it appears from the experimental evidence gathered so far that the concept of a fully determined pathway with additional small fluctuations is applicable to co-elastic crystals (Marais et al. 1991). The more important question concerns the nature of the Gibbs free enrgy G_{kin} which determines the pathway. Standard rate theories of Ising spin systems (Glauber. 1963; Kawasaki. 1966) show that G_{kin} is substantially different from the excess Gibbs free energy G discussed so far in this book for the case of thermodynamic equilibrium. Only at high temperatures ($T \geq T_c$), Dattagupta et al. (1991) found $G \sim G_{kin}$. For displacive systems and spin systems with a large number of local states , Marais et al. (1991) have shown both analytically and by computer simulation that the kinetic Gibbs free energy is, indeed, well approximated by the excess Gibbs free energy, i.e. $G_{kin} \sim G$. We will now sketch the line of argument using this approximation.

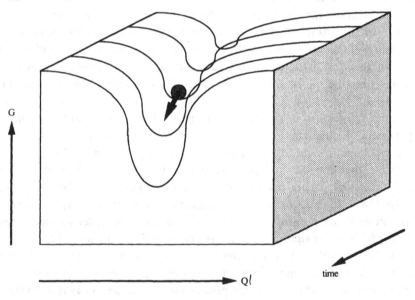

Fig.12.1 Potential valley $\Delta G(Q,t)$ in which the system follows the most probable pathway towards equilibrium. The slopes of the valley are steep so that fluctuations occur only on the bottom of the valley.

The starting point for the time-dependent order parameter theory is the Markovian master equation. Over a given small volume of crystal there is a probability, P, that some physical parameter will have a value n at time t. This probability will change with time according to the statistics of small changes, of magnitude a, occurring in the value of the local parameter n as:

$$\frac{1}{\gamma} \cdot \frac{\partial P(n,t)}{\partial t} = -\sum_a W(n \rightarrow n+a) P(n,t) + \sum_a W(n-a \rightarrow n) P(n-a,t)$$

12.1

(Glauber 1963; Kawasaki 1966; Salje 1988a). $W(n \rightarrow n+a)$ is the probability that the local parameter will jump from n to n+a in a unit of time given by $1/\gamma$, where γ is a characteristic frequency of the system. $W(n-a \rightarrow n)$ is the probability for the reverse, jump etc. It is then assumed that the jump probabilities follow Boltzmann statistics and that the most probable macroscopic pathway is given by integration of the most probable microscopic steps along the reaction pathway. In the macroscopic solution given by Salje (1988a), Dattagupta et al. (1991a,b) and Marais and Salje (1991), Q is eventually identified with n giving:

$$\frac{dQ}{dt} = \frac{-\gamma \langle a^2 \rangle}{2RT} \left(1 - \frac{\sinh(\xi \nabla)}{\xi \nabla} \right) \frac{\partial G}{\partial Q}$$

12.2

where R is the gas constant and ∇ is the gradient operator. The term $\langle a^2 \rangle$ is a correlation function which should be regarded as a material constant. The original physical significance of the 'a' parameter is lost in the integration procedure but, since it is related to the jump size for a given unit of time, it can be used to account for the temperature dependence of individual steps. The size of the jump in a fixed time at high temperatures will be greater than at low temperatures for an activated state model. Thus:

$$\langle a^2 \rangle = \lambda \exp\left(-\Delta G^*/RT\right)$$

12.3

where λ is a material constant and ΔG^* is the free energy of activation.

The term ξ is introduced in the statistical analysis in order to define the extent to which individual steps in the order parameter locally in a crystal will influence or induce changes in the same parameter elsewhere in the crystal. Formally, ξ is the correlation length of individual microscopic steps and it plays a crucial role in determining the overall kinetic behaviour.

Salje (1988a) finally rescaled equation 12.2 by introducing a second important length parameter, ξ_c, which represents the local length scale over which conservation occurs. The rate equation proposed by him combines all relevant cases in a universal way. It can be written as (Marais and Salje 1991):

$$\frac{dQ}{dt} = \frac{-\gamma \langle a^2 \rangle}{2RT} \left(1 - \frac{\xi_c^2}{\xi^2} \frac{\sinh(\xi V)}{\xi V}\right) \frac{\partial G}{\partial Q} \propto \frac{-(\tau_1^{-1} + \tau_2^{-1})}{RT} \left(1 - \frac{\tau_2^{-1}}{\tau_1^{-1} + \tau_2^{-1}} \frac{\sinh(\xi V)}{\xi V}\right) \frac{\partial G}{\partial Q}$$

12.4

where $\tau 1$ and $\tau 2$ are the characteristic times for the non-conserved and conserved process respectively.

In ferroelastic and co-elastic materials we assume that the correlation length ξ is induced by the stress fields and extends over several hundred Ångstroms. The second length scale of the conservation law is most easily explained in an example. Take, first of all, a spinodal exsolution process. An increase of the order parameter Q at a given position in the spinodal modulation will be compensated by an equivalent decrease in adjacent areas. The smallest possible length scale over which such changes compensate for each other is the length scale of the modulation itself, typically 100-1000Å. Defect related modulations in ferroelastics may show similar correlations and we find that the two length scales ξ and ξ_c are of similar orders of magnitude. In an order/disorder phase transition, on the other hand, we find that the net balance of the exchange process is only between neighbouring sites in the structure and ξ_c is of the order of interatomic distances. In purely displacive phase transitions, the order parameter is not conserved at all and the length ξ_c is strictly zero. In summary, we find that the parameter ξ_c/ξ is close to unity in spinodal processes, rather small for order/disorder transitions and zero for displacive systems. All three cases can occur in ferroelastic materials with coupling to sluggish order parameters.

An essential implication of the rate law in equation 12.4. is that for all finite values of ξ_c we expect structural modulations to occur even if they do not represent equilibrium states. This striking conclusion follows simply from the fact that a gradient operator appears in the rate law leading to a wave equation:

$$\frac{dQ}{dt} \propto \nabla^2 Q + \ldots$$

12.5

Structural modulations and, more general, non-uniform distributions of the order parameter are thus a rather typical feature of the non-equilibrium state during a kinetic experiment. Typical examples of the microstructures of kinetically disordered Na-feldspar and Mg-cordierite are shown in Fig.12.2 and 11.16.

We now discuss some specific features of the non-uniform kinetic states. Here it is sufficient to envisage the modulations as random fluctuations, ignoring their internal degree of organisation (see Chapter 11). The thermodynamically induced fluctuations are related to the order parameter susceptibility via the dissipation-fluctuation theorem with the imaginary part of the generalized susceptibility $\chi(\omega)$ (5.15; Landau and Lifshitz 1980):

$$\langle \delta Q^2 \rangle = \frac{\hbar}{p} \int_0^\infty \chi(\omega) \coth \frac{\hbar \omega}{2kT} d\omega \qquad\qquad 12.6$$

which becomes in the high temperature approximation for:

$$\chi(\omega) = \chi \, \delta(\omega - \omega_0), \quad kT \gg \hbar \omega_0 :$$

$$\langle \delta Q^2 \rangle = kT \cdot \chi \qquad\qquad 12.7$$

Strain interaction normally limits the wavevectors along which these fluctuations may occur along the well-defined directions. Restoring forces acting against the fluctuations are again due to the positive gradient energy

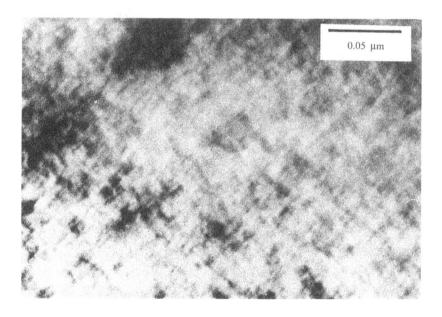

Fig.12.2 Microstructure of kinetically disordered Na-feldspar. The starting material and the final product are completely uniform crystals without the tweed structure (Courtesy of A. Putnis, Cambridge).

and the positive curvature of the Landau potential around the equilibrium point. The first contribution, namely the gradient force, also exists in the kinetic state and limits the amplitude of the fluctuation. Materials with very large, positive gradient energies will tend not to develop modulated microstructures. The second contribution due to the curvature of the Landau potential, on the other hand, does not exist in the kinetic state because this state does not represent an equilibrium point. This is clearly seen in the potential diagrams in Fig.12.3. Whereas the equilibrium points I, III, VI are in a minimum of the potential, the kinetic states along the dotted lines (or, at an early state, II) are situated on the slope of the potential. Large fluctuations in these states are actually energetically favoured if they lead towards the equilibrium state. They provide a means for the system to reduce its energy

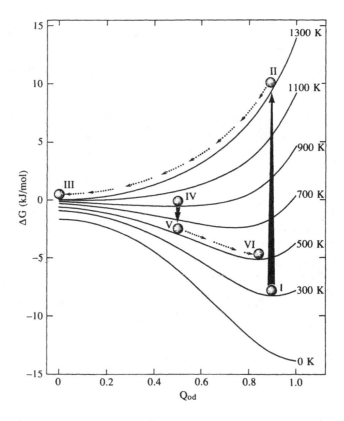

Fig.12.3 Excess Gibbs free energies of Na-feldspar for various temperatures as a function of the degree of Al,Si ordering. Kinetic pathways include shock-heating (I→II) and relaxation (II→III). A quench experiment includes the drop in temperature (IV→V) and the relaxation into equilibrium (V→VI).

and will lead to nucleation of the equilibrium phase if this nucleation process is not suppressed by large positive wall energies (or gradient energies). Nucleation and growth of large clusters is thus the other extreme of a kinetic process which occurs for small gradient energies. This process has been discussed in some detail in Chapter 8.1. In this chapter we are concerned with a kinetic mechanism in which the equilibrium state is approached in a more continuous manner with local fluctuations involving strain modulations but no classical nucleation.

12.1 Solution of the kinetic rate law in the Cahn (C) and the Ginzburg-Landau (GL) limit

A quantitative rate law can be obtained from the general rate law in equation 12.2 if the driving force $\partial G/\partial Q$ is known. In the case of fully conserved order parameters, the solutions have been discussed in great detail within the context of spinodal decomposition. Langer (1971) showed that a general solution of the type 12.4 can be reduced to the more familiar diffusion equation of Cahn with a simple parabolic Gibbs free energy. The Cahn expression results directly from equation 12.2. In the lowest order approximation:

$$\frac{dc}{dt} = d\nabla^2 c + \frac{D}{g''} 2\eta^2 Y \nabla^2 c - \frac{\chi 2\kappa D \nabla^4 c}{g'} \qquad 12.8$$

where the order parameter is proportional to the concentration c of a given species, D is an effective interdiffusion coefficient, η is the linear strain per unit volume, Y is a function of the elastic constants , χ is the susceptibility and κ is traditionally denoted as the gradient coefficient (Hilliard 1970). We follow here the usual convention and distinguish between the mainly elastic gradient interactions in equation 11.1 with the coefficients g and g' and the mainly 'chemical ' gradient term with the coefficient κ. Solutions are of the form:

$$C(k,t) = C_o(k,O)e^{\alpha(k)t} \qquad 12.9$$

with the amplification factor, α k), given by:

$$\alpha(k) = -D|k|^2 \left(1 + \frac{2\eta^2 Y}{g''} + \frac{2\kappa}{g''} \cdot |k|^2 \right) \qquad 12.10$$

Now, since it is possible to describe ordered structures in terms of composition modulations with wavelengths which are some fraction of the unit cell size, equations 12.8 and 12.9 might be used as a basis for treating the kinetics of ordering or disordering. For exsolution processes, steep concentration gradients are energetically unfavourable ($\kappa > 0$) and g" is negative. In systems with g">0 and $\kappa < 0$, steep concentration gradients will be energetically favourable and ordering may occur. Because of the simplifications introduced into the modified diffusion equation, particularly in linearising the gradient term and the local energy term (g" is assumed to be constant), however, the rate law strictly applies only for wavelengths which are considerably larger than the atomic spacings and for small amplitudes. Cook et al. (1969) were nevertheless able to derive a discrete solution for short wavelengths in a cubic lattice which required a relatively minor modification of the original continuum expression to include the contributions of higher order terms in k; $|k|^2$ is replaced by $B^2(k)$ where:

$$B^2(k) = \sum_r m(r)\left[1 - \cos\left(\overline{k} \cdot \overline{x}(r)\right)\right]$$

$$12.11$$

The term $m(r)$ takes into account different jump frequencies for atoms moving into different neighbouring sites and r is the nearest neighbour site reached by vector $x(r)$. Further adjustments can be made to the elastic energy to account for atomic scale anisotropies in elastic energies, or to the gradient energy in order to include the effects of higher order terms, but these are increasingly complex.

From an experimental point of view, the intensity of a satellite reflection, $I(k,t)$, is proportional to the square of the amplitude of the composition modulation, $A(k,t)$, and it is expected that:

$$I(k,t) = I(k,o)\exp(2\alpha(k)\,t)$$

$$12.12$$

A plot of $\ln(I(k,t)/I(k,o))$ against t for the decay of a modulated phase should therefore be linear and, for a disordering reaction this would give $\ln(Q^2) \propto t$.

Ferroelastics and co-elastics often relate their gradient energies to the formation of local strain fields as discussed in Chapter 11. Let us now consider this case and ignore all conservation of the order parameter (i.e. $\xi_c = 0$) The kinetic equation 12.2 reduces then to:

$$\frac{dQ}{dt} = -\frac{\gamma\lambda}{2kT}\,e^{-\Delta G^*/kT}\,\frac{\partial G}{\partial Q} = -\frac{1}{\tau kT}\,e^{-\Delta G^*/kT}\,\frac{\partial G}{\partial Q}$$

$$12.13$$

The relevant Gibbs free energy G is the relaxed excess Gibbs free energy of the structural phase transition. A typical example is shown in Fig.12.3. Here the excess Gibbs free energy of albite has been calculated from experimental data using the condition that the fast displacive order parameter Q has relaxed into the partial minimum for each value of the slow order parameter Q_{od}. The rate law is then determined by the evolution of the slow order parameter Q_{od} alone. A typical course of a kinetic experiment is then the shock-heating of ordered material (point I in the diagram) to high temperatures above T_c (e.g. point II). The crystal is then still ordered but the fast variables such as the strain, the optical phonons etc. have accomodated the temperature change. Annealing the crystal at the high temperature leads to a slow variation of the degree of ordering (along the dotted line in the diagram) until the total equilibrium is reached (point III). Another kinetic pathway involving ordering rather than disordering is also shown (IV-V-VI). We can see immediately that the Gibbs free energies are very different for the two cases and that, consequently, the time dependencies will be rather different. Let us now consider the two most simple rate laws following from 12.12.

a A homogeneous crystal and a second order phase transition

If the crystal undergoes a second order phase transition with a single, scalar order parameter it might well be that the total excess energy is the excess energy of the transition itself. We can substitute:

$$G = \frac{1}{2} A(T-T_c) Q^2 + \frac{1}{4} B Q^4 \qquad\qquad 12.14$$

into the equation 12.13 and find:

$$\frac{dQ}{dt} = - \frac{\gamma\lambda}{2kT} e^{-\Delta G^*/kT} \left[A(T-T_c) Q + B Q^3 \right] \qquad\qquad 12.15$$

or

$$t-t_0 = \int_{Q_o}^{Q} - 2kT/ \left(\gamma\lambda\, e^{-\Delta G^*/kT} \right) \frac{dQ}{A(T-T_c) Q + B Q^3} \qquad\qquad 12.16$$

where ΔG^* may be approximated by $\Delta G^*_0 + \varepsilon Q^2$. Ignoring this small Q dependence of ΔG^* we can solve the integral numerically. For $T \gg T_c$ we find:

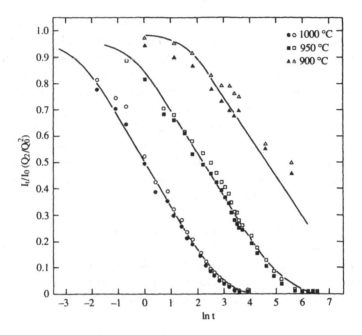

Fig.12.4 Time dependence of X-ray intensities as a function of the square of the order parameter in omphacite (Carpenter et al. 1989). The solid lines show the calculated rate laws using a 2-6 potential in equation 12.16.

$$t - t_o = - \cfrac{2kT}{\left[\gamma\lambda \exp\left(\dfrac{\Delta G^*}{kT}\right) A(T-T_c) \right]} \ \ln\left(\frac{Q}{Q_o}\right) \qquad\qquad 12.17$$

or:

$$Q \propto \exp - (t - t_0)/\tau \qquad\qquad 12.18$$

A typical example for this rate law is a homogeneous cation disorder at temperatures which are high compared with the transition temperature. Experimental results for the cation ordering in omphacite are shown in Fig.12.4. Solutions for three different temperatures are also plotted in Fig.12.4. It is obvious that tricritical and first order phase transitions can be treated in essentially the same way leading to explicit rate laws for specific coefficients of the Landau potential. We refer the reader to Salje (1988) and Carpenter and Salje (1989) for specific examples.

b 'Nonconvergent' ordering

The rate equations discussed so far are independent of whether or not a phase transition actually occurs. One of the most prominent cases is the so-called 'nonconvergent' ordering in which the order parameter never disappears and where, therefore, no high-symmetry phase exists. It appears that in some of these cases, the ordering process is independent of any symmetry relationships and we may thus write the excess Gibbs free energy without symmetry constraints:

$$G = HQ + \frac{1}{2} A' Q^2 + \frac{1}{3} B Q^3 + \dots \qquad 12.19$$

The linear term has then the usual meaning of a field term. All coefficients can now be temperature dependent because there is no underlying transition mechanism which might lead to further constraints. The inverse rate law is then for homogeneous systems:

$$\frac{\Delta t}{\tau} = - \exp\left(\Delta G^*/kT\right) \int \left\{- H + A' Q + BQ^2\right\} dQ \qquad 12.20$$

This rate law is, in particular, useful for the description of chemical processes which do not involve symmetry constraints. In order to show the close relationship with the rate law of chemical reactions, let us start from the reaction equation (Mueller 1967):

$$- \frac{\partial X_A^a}{\partial t} = \vec{k}\, X_A^a X_B^b - \overset{\leftarrow}{k}\, X_A^b X_B^a \qquad 12.21$$

where X_A^a is the occupancy of species A on site a, X_B^a is the occupancy of species B on site a, etc. The condition of a steady state is then:

$$\vec{k} = \overset{\leftarrow}{k}\ K_D^{-1} \qquad 12.22$$

with a constant factor K_D. We can now use the condition of the conservation of chemical species $X_A^a + X_A^b = C_A$, $X_B^a + X_B^b = C_B$ where C_A and C_B are the total concentration of A and B atoms, respectively. The reaction-equation 12.20 can then be cast into the form:

$$- \frac{\partial X_A^a}{\partial t} = \vec{k}\left(a + b\, X_A^a + c \left(X_A^a\right)^2\right) \qquad 12.23$$

We can now express the occupancy X_A^a as a function of the order parameter Q. If, for example, the atom A can be distributed over n sites, one of which is the 'correct' site a, we find:

$$X_A^a = \frac{1}{n}\,(Q+1)$$

$$\text{12.24}$$

Expressing X_A^a by Q in equation 12.22 leads to the rate law 12.19 with:

$$H = -n\,\vec{k}\left(a + \frac{b}{n} + \frac{c}{n^2}\right)$$

$$\text{12.25}$$

$$A' = \vec{k}\left(\frac{2c}{n} - b\right)$$

$$\text{12.26}$$

$$B = -\frac{c\vec{k}}{n}$$

$$\text{12.27}$$

A typical example for this behaviour is the non-convergent ordering in sanidine with n=4 (Salje and Kroll 1990).

12.2 Strain modulations and the effect of local fields

The effect of fluctuations does not only apply to (partly) conserved order parameters but also plays an important role for non-conserved order parameters. This effect becomes obvious if we extend the Gibbs free energy to include the Ginzburg energy as introduced in Chapter 11. Taking only the quadratic term into account, we find for a second order phase transition:

$$\frac{dQ}{dt} = -\frac{\Delta Q}{\tau} - \frac{\gamma\lambda}{2kT}\,\exp\left(\frac{-\Delta G^*}{kT}\right)\,g\,\nabla^2 Q$$

$$\text{12.28}$$

where the time constant τ is given by:

$$\tau \approx \frac{kT}{\left[\gamma\lambda\,\exp\left(\frac{-\Delta G^*}{kT}\right)b\,Q_{eq}^2\right]}$$

$$\text{12.29}$$

We can now treat Q again as the sum of Fourier components with:

$$Q = \Sigma\,Q_k\,e^{ikr}$$

$$\text{12.30}$$

and find the solution:

$$\ln\,\Delta Q \propto \frac{t-t_0}{\tau_k}$$

$$\text{12.31}$$

with:

$$\tau_k^{-1} = \tau^{-1} + \frac{\gamma\lambda}{kT} e^{\frac{-\Delta G^*}{kT}} g\, k^2 \qquad\qquad 12.32$$

The time constant is thus explicitly dependent on the wavevector of the fluctuation. We see that the fastest relaxations occur for large k-values and positive g-values because the gradient energy acts as an additional driving force for the kinetic process. Changes of the order parameter will occur more slowly for fluctuations with long wavelengths (i.e. small k-vectors) and we expect that the fluctuation spectrum after some time will be dominated by long wavelength modulations. The time constant τ relates to the finite relaxation time of the slowest component at k=0. The relaxation spectrum of Q_k is then:

$$Q_k \propto \exp\left\{-(t-t_0)\left(x + y\, k^2\right)\right\} \qquad\qquad 12.33$$

where $x = \tau^{-1}$ and $y = \frac{\gamma\lambda g}{kT}\exp\left(\frac{-\Delta G^*}{kT}\right)$ \qquad 12.34

This result shows that Q_k has a Gaussian distribution around k=0 with the width $1/\, y(t-t_0)$. This width decreases rapidly with time from the uniform distribution of a white noise spectrum at $t=t_0$ to an infinitely narrow delta function at infinite time.

Local nucleation processes are much influenced by this distribution function. Let us consider an ordering process. If the critical point for the ordering happens to be at the zone centre, many fluctuations with the 'correct' wavevector k=0 occur and the distribution will sharpen up around the critical point. The distribution function will show fluctuation-induced nucleation (Salje 1988) with a spread of k-vectors around the critical point. As the fluctuations will display the same general features as discussed in Chapter 11, we expect irregular soliton lattices to occur as the most probable microstructure.

This behaviour can now be contrasted with that of the zone boundary instabilities. Firstly, the spontaneous strain is now a quadratic function of the order parameter and the strain gradients will constrain the elastic fluctuations in the same way as for proper ferroelastic phase transitions. The maximum of their fluctuation distribution is still at the zone centre whereas the order parameter leads to structural instabilities at the zone boundary. Ordering can now be achieved by only one Fourier component which is not supported by the strain fluctuations. We may thus expect that strain fluctuations are, in general, less relevant at zone boundary instabilities than for phase transitions with instabilities at the zone centre.

The role of the fluctuations for the kinetic process can be amplified by impurities and long-lived modulations. If the local environments, such as discussed in Chapter 9.3 exert a significant influence on the kinetics in a given region in the crystal, the kinetic rate law in equation 12.4 has to include such local fields explicitly. Salje (1988) following earlier arguments by Bray and Moore (1982) has shown that on a local scale the correlation function:

$$\sigma = \langle Q^2 \rangle \qquad\qquad 12.35$$

follows in mean field approximation the rate law:

$$\ln \sigma \propto t \left(1 - \tanh\left[\frac{M_{eff}}{kT}\right]\right) \qquad\qquad 12.36$$

where M_{eff} is the effective field seen by the order parameter. This field can now fluctuate and the probability of finding a given local field will depend on the probability of finding the corresponding configuration of the Q. For a Gaussian distribution of M_{eff} integration of equation 12.36 leads to a rate law of the average value of:

$$\langle \sigma \rangle \approx A k T \ln t + B \qquad\qquad 12.37$$

where A and B are constants at constant temperature.

The physical meaning of these random fields in the context of the kinetic behaviour of ferroelastic and co-elastic crystals can best be explained in the case of the harmonic approximation of the Landau potential by:

$$\frac{\partial G}{\partial Q} = \chi_{eff}^{-1} Q \qquad\qquad 12.38$$

The effective susceptibility may then consist of two parts, namely the intrinsic order parameter susceptibility and the contributions from the fields:

$$\chi_{eff}^{-1} = \chi_h^{-1} + \chi_r^{-1} \qquad\qquad 12.39$$

The time evolution of the order parameter depends then on the probability distribution of χ^{-1}_{eff}. This distribution is called $P(\chi^{-1}_{eff})$ and the general rate law can now be written as a simple Laplace transform of this distribution function (Salje and Wruck 1988):

$$\langle Q \rangle \propto \int P(x) \, e^{-xt} \, dx \qquad\qquad 12.40$$

where the integration parameter is $x = \chi^{-1}_{eff}$.

The probability function P can now be determined from experimental observations by inverse Laplace transformation. The following five models for P are of particular interest because they seem to best describe the experimental data (see Salje and Wruck 1988 for details):

a Polynomial distribution of $P(x)$ leads to Taylor expansions of $\langle Q \rangle$ as:

$$\langle Q \rangle = \frac{\rho_1}{t} + \frac{\rho_2}{t^2} + \dots$$

where $\rho_1, \rho_2 \dots$ are the coefficients of the polynomial form of P.

b Gaussian distributions of P lead to a rate law of:

$$\langle Q \rangle = e^{-x_o t} \, \exp\left(\tfrac{1}{2}\Gamma t^2\right) \, \mathrm{erfc}\left(\sqrt{\tfrac{\Gamma}{2}} \, t\right)$$

where x_o is the most probable inverse time constant, Γ is the Gaussian line width and erfc is the complement error function.

c A Maxwell distribution of P leads to the simple rate law:

$$\langle Q \rangle \propto e^{-k\sqrt{t}}$$

d Pseudo glasses possess a logarithmic rate law:

$$\langle Q \rangle \propto \ln t$$

e Power laws with $P(x) = x^\alpha$ lead to rate laws of:

$$\ln \langle Q \rangle = -(\alpha + 1) \ln t$$

with $\alpha > 1$.

Only a few systematic experimental investigations of the time dependence of ferroelastic and co-elastic phase transitions have yet been undertaken and it is expected that much research activities will focus on these subjects in future.

References

ACHIAM, Y. & IMRY, Y. (1975). Phase transitions in systems with a coupling to a non-ordering parameter. *Phys. Rev.* **B12**, 2768.

ADLHART, W., FREY, F. & JAGODZINSKI, H. (1980) X-ray and neutron investigations of the $P\bar{1}$-$I\bar{1}$ transition in pure anorthite. *Acta Cryst.* **A36**, 450.

AIZU, K. (1970). Determination of state parameters and formulation of spontaneous strain for ferroelastics. *J. Phys. Jap.* **28(3)**, 706.

AIZU, K. (1973). Second-order ferroic state shifts. *J. Phys. Jap.* **34(1)**, 121.

AIZU, K. (1976). Interpretations of sequence of transitions in barium sodium niobate $Ba_2NaNb_5O_{15}$. *J. Phys. Jap.* **41(3)**, 880.

ALEKSANDROV, K.S. & FLEROV, I.N. (1979). (RS) Applicable region of thermodynamical theory for structural phase transitions close to tricritical point. *Fiz. Tverd T.* **21(2)**, 327.

ALEKSANDROV, K.S., BOVINA, A.F., VORONOV, V.N., GOREV, M.V., ISKORNEV, I.M., MELNIKOVA, S.V., MISJUL, S.V., PROKERT, F. & FLEROV, I.N. (1985). Ferroelastic phase transitions in Elpasolites $A_2BB^{3+}X_6$. *Jap. J. Appl. Phys. Suppl.* **24-2**, 699.

AMELINCKX, S. & VAN LANDUYT, J. (1976) in H.-R. Wenk (Ed.): *Electron Microscopy in Mineralogy.* Springer: Berlin.

ANISIMOV, M.A., GORODETS, E.E., & ZAPRUDSK, V.M. (1981).Phase transitions with coupled order parameters-a review. *USP Fiz NAU* **133(1)**:103.

AUBRY, S. (1976). New interpretation of dynamical behaviour of a displacive model. *Ferroelectr.* **12(1-4)**, 263.

AUBRY, S. & LEDAERON, P.Y.(1983).The discrete Frenkel-Kontorova model and its extensions.1.Exact results for the ground states. *Physica D* **8(3)**, 381.

BAISA, D.F., BONDAR, A.V., GORDON, A.S., & MALTSEV, S.V. (1979). Spin relaxation in first-order phase transitions and in the neighbourhood of the critical point. *Phys. Stat. Sol.* **93**, 805.

BAK, P. & EMERY, V.J. (1976). Theory of the structural phase transformations in tetrathia-fulvalene-tetracyanoquinodimethane (TTF-TCNQ). *Phys. Rev. Lett.* **36(16)**, 978.

BALAGUROV, A.M., POPA, N.C., & SAVENKO, B.N. (1986). Symmetry analysis of the low-temperature domain structure in ferroelastic $LiKSO_4$. *Phys. Stat. Sol. b* **134**, 457.

BARSCH, G.R. & KRUMHANSL, J.A. (1984) Twin boundaries in ferroelastic media without interface dislocations. *Phys. Rev. Lett.* **53**, 1069.

BARSCH, G.R. & KRUMHANSL, J.A. (1988). Nonlinear and nonlocal continuum model of transformation precursors in martensites. *Met. Trans. A.* **19A**, 761.

BARTEL, J.J. & WESTRUM, E.F. (1976). Thermodynamics of Fe(II) Fe(III) oxide systems 2. Zinc doped and cadmium doped Fe_3O_4 and crystalline magnetite. *J. Chem. Ther.* **8(6)**, 583.

BASTIE, P., LAJZEROWICZ, J. & SCHNEIDER, J.R. (1978) Investigation of the ferroelectric-ferroelastic phase transition in KH_2PO_4 and RbH_2PO_4 by means of γ-ray diffractometry. *J. Phys. C.* **11**, 1203.

BENGUIGUI, L. & BEAUCAMPS, Y. (1980). A new tricritical point in Co doped barium-titanate. *Ferroelectr.* **25(1-4)**, 633.

BIRMANN, J.L. (1966). Full group and subgroup methods in crystal physics. *Phys. Rev.* **150**, 771.

BISMAYER, U., SALJE, E. & JOFFRIN, C. (1982). Reinvestigation of the stepwise character of the ferroelastic phase transition in lead phosphate-arsenate, $Pb_3(PO_4)_2$-$Pb_3(AsO_4)_2$. *J. Physique* **43**, 1379.

BISMAYER, U., SALJE, E., GLAZER, A.M. & COSIER, J. (1986). Effect of strain-induced order parameter coupling on the ferroelastic behaviour in lead phosphate-arsenate. *Phase Trans.* **6**, 129.

BISMAYER, U. & SALJE, E. (1981). Ferroelastic phases in $Pb_3(PO_4)_2$-$Pb_3(ASO_4)_2$ X-ray and optical experiments. *Acta Cryst.* **A37**, 145.

BISMAYER, U., SALJE, E., JANSEN, M. & DREHER, S. (1986). Raman scattering near the structural phase transition of As_2O_5: order parameter treatment. *J. Phys. C.* **19**, 4537.

BLINC, R., PRELOVSEK, P. & KIND, R. (1983). Amplitude-fluctuation effects on magnetic-resonance line shape and soliton density in incommensurate systems. *Phys. Rev.* **B27**, 5404.

BRAY, A.J. & MOORE, M.A. (1982).Spin glasses:the hole story.*J.Phys. C* **15**, 2417.

BRODY, E.M. & CUMMINS, H.Z. (1974). Brillouin scattering study of elastic anomaly in ferroelectric KH_2PO_4. *Phys. Rev. B.* **9**, 179.

BRUCE, A.D. & COWLEY, R.A. (1981). *Structural phase transitions*. London: Taylor and Francis.

BURGERS, W.G. (1945).The genesis of twin crystals. *Am. Min.* **30**, 469.

CARPENTER, M.A., McCONNELL, J.D.C. & NAVROTSKY, A. (1985). Enthalpies of ordering in the plagioclase feldspar solid solution. *Geochim. Cosmochim. Acta* **49**, 947.

CARPENTER, M.A. (1988). Thermochemistry of aluminium/silicon ordering in feldspar minerals. In *Physical properties and thermodynamic behaviour of minerals* (Ed. E. Salje). Dordrecht: Reidel.

CARPENTER, M.A. & SALJE, E. (1989). Time-dependent Landau theory for order/disorder processes in minerals. *Min. Mag.* **53**, 483.

CARPENTER, M.A., DOMENEGHETTI, M.C. & TAZZOLI, V. (1990). Application of Landau theory to cation ordering in omphacite II, kinetic behaviour. *Europ. J. Mineralogie* **2**, 19.

CHAN, S-K. (1977). Steady-state kinetics of diffusionless first order phase transformations. *J. Chem. Phys.* **67(12)**, 5755.

CLAPP, P.C., RIFKIN, J., KENYON, J. & TANNER, L.E. (1988). Computer study of tweed as a precursor to a martensitic transformation of a bcc lattice. *Metall. T-A* **19(4)**, 783.

CLARKE, R. & GLAZER, A.M. (1976). Critical phenomena in ferroelectric crystals of lead zirconate titanate. *Ferroelectr.* **14**, 695.

COOK, H.E. (1969).The kinetics of clustering and short range order in stable solid solutions *J Phys Chem. Solids.***30**, 2427.

COLLINS, M.A., BLUMEN, A. CURRIE, J.F. ROSS, J. (1979). Dynamics of domain walls in ferrodistortive materials I - theory. *Phys. Rev.* **B19**, 3630.

COWLEY, R.A. (1976). Acoustic phonon instabilities and structural phase transitions. *Phys. Rev.* **B13**, 4877.

CRACKNEL, A.P. (1974). Group theory in solid state physics is not dead yet alias some recent developments in use of group theory in solid state physics. *Adv. Phys.* **23(5)**, 673.

DATTAGUPTA, S., HEINE, V., MARAIS, S. & SALJE, E. (1991). Rate equation for atomic ordering in mean field theory I: uniform case *J.Phys.Cond.Matt.* **3**, 2963.

DATTAGUPTA, S., HEINE, V., MARAIS, S. & SALJE, E. (1991). Rate equation for atomic ordering in mean field theory II: general considerations *J.Phys.Cond.Matt.* **3**, 2975.

DAVID, W.I.F. (1983). Structural relationships between spontaneous strain and acoustic properties in ferroelastics. *J. Phys. C.* **16**, 2455.

DEVARAJAN, V. & SALJE, E. (1984). Phase transition in $K_2Cd_2(SO_4)_3$: investigation of the non-linear dependence of spontaneous strain and morphic birefringence on order parameter as determined from excess measurements. *J. Phys. C* **17**, 5525.

DMITRIEV, V.P., ROCHAL, S.B., GUFAN, YU.M. & TOLEDANO, P. (1989). Reconstructive transitions between ordered phases: the martensitic fcc-hcp and the graphite-diamond transitions. *Phys. Rev. Lett.* **62**, 2495.

DOVE, M.T. & POWELL, B.M. (1989).Neutron diffraction study of the tricritical orientational order/disorder phase transition in calcite at 1260K.*Phys. Chem. Min.* **16**, 503.

DUBROVSKY, I.M. & KRIVOGLAZ, M.A. (1979). (RS) Phase transitions of the 2nd kind in crystals containing dislocations. *Zh. EKSPTEO* **77(3)**: 1017.

DURAND et al 1985. As quoted by Denoyer, F. & Currat, R. in 'Incommensurate phases in dielectrics materials'. (Ed. Blinc & Levanyuk 1986). North Holland.

DVORAK, V. (1972). Structural phase-transitions in langbeinites. *Phys. Stat. Sol.-B.* **52(1)**, 93.

DVORAK, V. (1978). Structural phase-transitions from a hypothetical phase. *Czec. J. Phys.* **28(9)**, 989.

ELLIOTT, R.J.,YOUNG, A.P.,& SMITH, S.R.P. (1971). Acoustic anomalies in Jahn-Teller coupled systems..*J Phys C.* **4**, L317.

FALK, F. (1983). Ginzburg-Landau theory of static domain walls in shape memory alloys. *Z Phys B -Condensed Matter* **51**, 177.See also:

FALK, F. (1982). Landau theory and martensitic phase transitions. *J. de Physique* **43(12)**, C4-3.

FEILE, R., LOIDL, A., & KNORR, K. (1982). Elastic properties of $(KBr)_{1-x}(KCN)_x$. *Phys. Rev.* **B26**, 6875.

FLOCKEN, J.W., GUENTHER, R.A., HARDY, J.R. & BOYER, L.L. (1985). First-principles study of structural instabilities in halide-based perovskites: competition between ferroelectricity and ferroelasticity. *Phys. Rev.* **B31**, 7252.

FOLK, R., IRO, H. & SCHWABL, F. (1979). Critical dynamics of elastic phase transitions.*Phys.Rev. B* **20(3)**, 1229.

FOUSEK, J. & JANOVEC, V. (1969). Orientation of domain walls in twinned ferroelectric crystals. *J. Appl. Phys.* **40**, 135.

FUCHIZAKI, K. & YAMADA, Y. (1989).Anomalous incommensurability and embryonic fluctuations at first order phase transformations.*Tech.Rep. of ISSP Ser A* , no 2116.

GARLAND, C.W. & NOVOTNY, D.B. (1969). Ultrasonic velocity and attenuation in KH_2PO_4.*Phys. Rev.* **177**, 971.

GARLAND, C.W. & BALOGA, J.D. (1977). Heat capacity of NH_4Cl and ND_4Cl single crystals. *B.Am. Phys. S.* **22(3)**, 299.

GEHRING, G.A. & GEHRING, K.A. (1975). Cooperative Jahn-Teller effects. *Rep. Dr. Phys.* **38(1)**, 1.

GHOSE, S. (1989).Personal communication.

GIDDY, A.P., DOVE, M.T. & HEINE, V. (1989). What do Landau free energies really look like for structural phase transitions? *J.Phys.Cond.Matt.* **1**, 8327.

GLAZER, A.M. (1989). Linear and circular birefringence and crystal structures. In *Thermodynamic behaviour and physical properties of minerals*. (Ed. E. Salje). Dordrecht: Reidel.

GLAUBER, R.J. (1963). Time-dependent statistics of the Ising model. *J. Math. Phys.* **4**, 294.

GMELIN, E. & ROEDHAMME, P. (1981). Automatic low-temperature calorimetry for the range 0.3-320K. *J. Phys. E.* **14(2)**, 223.

GOLDSMITH, J.R. & LAVES, F. (1954). On the superstructure of gallium containing anorthites and germanium-containing anorthites. *Acta Cryst.* **7(1)**, 131.

GORDON, A. (1983). Tricritical phase transitions in ferroelectrics. *Physica* **122B**, 321.

GRUNDY, H.D. & BROWN, W.L. (1969). A high temperature X-ray study of equilibrium forms of Albite.*Min.Mag.* **37**, 156.

GUFAN, YU.M. & LARIN, E.S. (1980). Theory of phase transitions described by two order parameters. *Sov. Phys. Sol. State* **22**, 270.

GURSKAS, A.A., ZVYAGIN, A.I. & PELIKH, L.N. (1981). Ferroelastic phase transitions in $CsR (MoO_4)_2$-$CsR(WO_4)_2$ with R=Dy, Ho, Er, Tm, Yb and Lu. *Ferroelectr.* **48(1-3)**, 81.

GURSKAS, A.A., KOKSHENER, V.B. & SYRKIN, E.S. (1987). Dynamical character of impurity influence on phase transitions in layered ferroelastics. *Ferroelectr.* **75**, 483.

HALPERIN, B.I. & VARMA, C.M. (1976). Defects and the central peak near structural phase transitions. *Phys. Rev. B* **14**, 4030.

HARRIS, M.J., SALJE, E., GUTTLER, B.K. & CARPENTER, M.A. (1989). Structural states of natural potassium feldspar: an infrared spectroscopic study. *Phys. Chem. Min.* **16**, 649..

HATTA, I., SHIROISH, Y., MULLER, K.A. & BERLINGER, W. (1977). Critical behaviour of heat capacity in $SrTiO_3$. *Phys. Rev.* **B16(3)**, 1138.

HATTA, I. & IKUSHIMA, A.J. (1981). Studies on phase-transitions by AC calorimetry. *JPN JA Phys.* **20(11)**, 1995.

HEINE, V. & McCONNELL, J.D.C. (1984). The origin of incommensurate structures in insulators. *J. Phys. C.* **17**, 1199.

HELWIG, J., PETERSSON, J. & SCHNEIDER, E. (1977). Landau behaviour of heat capacity of $AgNa(NO_2)_2$ close to nearly tricritical ferroelectric phase transition. *Z. Phys. B.* **28(2)**, 87.

HILLIARD, J.E. (1970). Spinodal decomposition, phase transformations (ed. H.I. Aaronson), *Am. Soc. Metals*, 497.

IIZUMI, M., AXE, J.D., SHIRANO, G. & SHIMAOKA, K. (1977). Structural phase transformation in K_2SeO_4. *Phys. Rev.* **B15**, 4392.

INTERNATIONAL TABLES FOR X-RAY CRYSTALLOGRAPHY (1985). Publ. Int. Union of Crystallography, Kynoch Press.

IMRY, Y. (1975). On the statical mechanics of coupled order parameters. *J. Phys. C.* **8**, 567.

ISHIBASHI, Y., HARA, K. & SAWADA, A. (1988). The ferroelastic transition in some Scheelite-type crystals. *Physica B* **150**, 258.

JACOBS, A.E. (1985). Solitons of the square-rectangular martensitic transformation. *Phys. Rev.* **B31(9)**, 5984.

JANOVEC, V. (1976). Symmetry approach to domain structures. *Ferroelectr.* **12(1-4)**, 43.

JANOVEC, V., DVORAK, V. & PETZELT, J. (1975). Symmetry classification and properties of equi-translation structure phase transformations. *Czech J Phys* **25(12)**, 1362.

JANSEN, M. (1977). Crystal structure of As_2O_5. *Angew Chem* **16**, 314.

JOFFRIN, C., BENOIT, J.P., CURRAT., R. & LAMBERT, M. (1979). Transition de phase ferroelastique du phosphate de plomb. Etude par diffusion inelastique des neutrons. *J. de Physique* **40**, 1185.

KALUS, C.K. (1978). Structural studies on pure anorthite $CaAl_2Si_2O_8$. *Acta Cryst A* **34(S)**, S29.

KEMPSTER, C.J., MEGAW, H.D. & RADOSLOV, E.W. (1962). Structure of Anorthite $CaAl_2Si_2O_8$ 1-Structure analysis. *Acta Crst* **15**, 1005.

KLEMAN, M. & SCHLENKE, M. (1972). Use of dislocation theory in magnetoelasticity. *J. Appl. Phys.* **43(7)**, 3184.

KNORR, K., LOIDL, A. & KYEMS, J.K. (1986). Ferroelastic transitions in KBr:KCN studied by neutrons, x-rays and ultrasonics. *Physica B* **136**, 311.

LANDAU, L.D. & KHALATNIKOV, I.M. (1954). *Dokl Akad Nauk SSR* **96**, 469.

LANDAU, L.D. & LIFSHITZ, E.M. (1980). *Statistical Physics*. Oxford: Pergamon Press.

LANGER, J.S. (1971). Theory of spinodal decomposition in alloys. *Ann. Phys.* **65**, 53.

LAJZEROWICZ, J. (1981). Domain wall near a 1st order phase transition - role of elastic forces. *Ferroelectrics* **35**, 219.

LAVES, F. (1969), personal communication.

LEVANYUK, A.P. (1963). In: *Statistical physics* (Ed. L.D. Landau & E.M. Lifshitz). Oxford: Pergamon Press, 482.

LEVANYUK, A.P. & SOBYANIN, A.A. (1970). Second order phase transitions without divergences in second derivatives of thermodynamic potential. *JETP Lett.* **11(11)**, 371.

LEVANYUK, A.P., OSIPOV, V.V., SIGOV, A.S. & SOBYANIN, A.A. (1979). Evolution of defect structure near the phase transition points and the anomalies in the properties of solids due to the defects. *ZH EKSP TEO* **76(1)**, 345.

LEVANYUK, A.P., SIGOV, A.S. & SOBYANYIN, A.A. (1980). The influence of defects on the properties of solids near phase transition points. *Ferroelectr.* **24**, 61.

LEVANYUK, A.P. & MINYUKOV, S.A. (1980). Influence of defects on properties of materials near phase transitions close to the tricritical point. *Sov. Phys. Solid State* **22(3)**, 507.

LINES, M.E. & GLASS, A.M. (1977). *Principles and applications of ferroelectrics and related materials*. Oxford: Clarendon Press.

LÜTI, B. & REHWALD, W. (1981). Ultrasonic studies near structural phase transitions. *Curr. Phys.* **23**, 131-184.

LYNDEN-BELL, R.M., FERRARIO, M., McDONALD, I.R. & SALJE, E. (1989). A molecular dynamics study of orientational disordering in crystalline sodium nitrate. *J. Phys. Condens. Matter* .**1**, 6523.

McCONNELL, J.D.C. (1965). Electron optical study of effects associated with partial inversion in a silicate phase. *Phil. Mag.* **11**, 1289.

McCONNELL, J.D.C. (1969). Electron optical study of incipient exsolution and inversion phenomena in the system $NaAlSi_3O_8$-$KAlSi_3O_8$. *Phil. Mag.* **19**, 221.

McCONNELL, J.D.C. (1971). Electron-optical study of phase transformations. *Min. Mag.* **38(293)**, 1.

McCONNELL, J.D.C. (1975). Microstructures of minerals as petrogenetic indicators. *Ann. Rev. Earth Plan. Sci.* **3**, 129.

MA, S.-K. (1976). *Modern theory of critical phenomena*. Reading, Mass.: Benjamin.

MAEDA, M. (1979). Elastic properties of $K_2Mn_2(SO_4)_3$. *J.Phys.Soc.Jap.* **47**, 1581.

MANOLIKAS, C., VAN TENDELOO, G. & AMELINCKX, S. (1986). The local structure of domain boundaries in ferroelastic lead orthovanidate. *Sol. State Comm.* **58**, 851.

MARAIS, S. & SALJE, E. (1991). Deviation of a rate law for non-uniform systems and continuous order parameters. *J.Phys.Cond.Matt.* **3**, 3677.

MARAIS, S., PADLEWSKI, S. & SALJE, E. (1991). On the origin of kinetic rate equations: Salje-Glauber-Kawasaki *J.Phys.Cond.Matt.* **3**, 6571.

MASHIYAMA, H. & TANISAKI, S. (1982). X-ray study on the modulation wave number in $N(CH_3)_42FeCl_4$. *J.Phys.C* **15**, L455.

MAYER, A.P. & COWLEY, R.A. (1988). The continuous melting transition of a three dimensional crystal at a planar elastic instability. *J. Phys. C.* **21**, 4827.

METIU, H., KITAHARA, K. & ROSS, J. (1976a). Stochastic theory of the kinetics of phase transitions. *J. Chem. Phys.* **64**, 292.

METIU, H., KITAHARA, K. & ROSS, J. (1976b). A derivation and comparison of two equations (Landau-Ginzburg and Cahn) for the kinetics of phase transitions. *J. Chem. Phys.* **65**, 393.

MUELLER, R.F. (1967). Model for order-disorder kinetics in certain quasi-binary crystals of continuously variable composition. *J. Phys. Chem. Solids* **28**, 2239.

MÜLLER, K.A., BERLINGER, W., & WALDNER, F. (1968). Characteristic structural phase transition in pervoskite-type compounds. *Phys. Rev. Lett.* **21**, 814.

MÜLLER, K.A., BERLINGER, W., WEST, C.H. & HELLER, P. (1974). Critical dynamics in $SrTiO_3$ from paramagnetic resonance. *Phys. Rev. Lett.* **32**, 160.

NORMAND, B.G.A., GIDDY, A.P., DOVE, M.T. & HEINE, V. (1990). Bifurcation behaviour in structural phase transitions with multi-well potentials. *J. Phys.*(in press).

NYE, J.F. (1964). *Physical Properties of Crystals*. Clarendon Press: Oxford.

OHNO, E. (1987). Theory of ferroelastic phase transition in NaClV, KClV and RbCN. *J. Phys. Soc. Jap.* **56**, 4414.

OLEKSY, C. & PRYSZTAWA, J. (1983). On phase transitions with bilinearly coupled order parameters. *Physica* **121A**, 145.

O'NEIL, M.J. (1966). Measurement of specific heat functions by differential scanning calorimetry. *Analyt. Chem.* **38**, 1331.

OVERHAUSER, A.W. (1971). Observability of charge-density waves by neutron diffraction. *Phys. Rev.* **B10(3)**, 173.

PALMER, D.C., PUTNIS, A. & SALJE, E. (1988). Twinning in tetragonal leucite. *Phys. Chem. Min.* **16**, 298.

PALMER, D.C., SALJE, E. & SCHMAHL, W.W. (1989). Phase transitions in leucite I: X-ray diffraction studies. *Phys. Chem. Min.* **16**, 606.

PARLINSKI, K. (1988). Molecular-dynamics simulation of incommensurate phases. *Comp. Phys. Rep.* **8(4)**, 157.

PEERCY, P.S. (1975). Raman Scattering near the tricritical point in SbSi. *Phys. Rev. Letters* **35**, 1581.

PERCIVAL, M.J.L. & SALJE, E. (1989). Optical absorption spectroscopy of the $P2_13$-$P2_12_12_1$ transformation in the $K_2Co_2 (SO_4)_3$ Langbeinite. *Phys. Chem. Min.* **16**, 563..

PERCIVAL, M.J.L., SCHMAHL, W.W. & SALJE, E. (1989). Structure of cobalt doped $K_2Cd_2(SO_4)_3$. Langbeinite at three temperatures above the $P2_13$-$P2_12_12_1$ phase transiton and a new trigger mechanism for the ferroelastic transformation. *Phys. Chem. Min.* **16**, 569.

POUGET, J. (1988). Nonlinear dynamics of lattice models for elastic media. In *Thermodynamic behaviour and physical properties of minerals* (Ed. E. Salje). Dordrecht: Reidel.

PUTNIS, A., REDFERN, S.A.T., FYFE, C.A. & STROBL, H. (1987). Structural states of Mg-Cordierite I: Order parameters from synchrotron, X-ray and NMR data. *Phys. Chem. Min.* **14(5)**, 446.

REEDER, R.J., REDFERN, S.A.T. & SALJE, E. (1988). Spontaneous strain at the structural phase transition in $NaNO_3$. *Phys. Chem. Min.* **15**, 605.

REDFERN, S.A.T., GRAEME BARBER, A. & SALJE, E. (1988). Thermodynamics of plagioclase III: Spontaneous strain at the $I\bar{1}$-$P\bar{1}$ phase transition in Ca-rich plagioclase. *Phys. Chem. Min.* **16**, 157.

REDFERN, S.A.T., SALJE, E. & NAVROTSKY, A. (1989). High temperature enthalpy at the orientational order-disorder transition in calcite: implications for the calcite/aragonite phase equilibrium. *Contrib. Min. Petrol.* **101**, 479.

REDFERN, S.A.T. & SALJE, E. (1987). Thermodynamics of plagioclase II: temperature evolution of the spontaneous strain at the $I\bar{1}$-$P\bar{1}$ phase transition in anorthite. *Phys. Chem. Min.* **14**, 189.

REDFERN, S.A.T. & SALJE, E. (1988). Spontaneous strain and the ferroelastic phase transition in As_2O_5. *J. Phys. C.* **21**, 277.

REDFERN, S.A.T. & SALJE, E. (1989). Thermodynamics of plagioclase II: temperature evolution of the spontaneous strain at the $I\bar{1}$-$P\bar{1}$ phase transition in anorthite. *Phys. Chem. Min.* **14**, 189.

REESE, W. (1969). Studies of phase transitions in order-disorder ferroelectrics 3 - Phase transition in KH_2PO_4 and a comparison with KD_2PO_4. *Phys. Rev.* **181**, 905.

REHWALD, W., RAYE, M., COHEN, R.W. & CODY, G.D. (1972). Elastic moduli and magnetic susceptibility of monocrystalline Nb_3Sn. *Phys.Rev. B* **6(2)**, 363.

REHWALD, W. (1973). The study of structural phase transitions by means of ultrasonic experiments. *Adv. Phys.* **22(6)**, 721.

REIMER, L. (1984). *Transmission electron microscopy*. Series in Optical Sciences Vol. 36. Berlin: Springer-Verlag.

ROCQUET, P. & COUZI, M. (1985). Structural phase transition in chiolite $Na_5Al_3F_{14}$: II - A model for the pseudo-proper ferroelastic transition. *J. Phys. C.* **18**, 6571.

ROCQUET, P., COUZI, M., TRESSAUD, A., CHAMINADE, J.P. & HAUW, C. (1985). Structural phase transition in chiolite $Na_5Al_3F_{14}$ 1. Raman Scattering and X-ray diffraction study. *J. Phys. C.* **18(36)**, 6555.

ROSENCWAIG, A. & GERHO, A. (1976). Theory of photoacoustic effect with solids. *J. Appl. Phys.* **47(1)**, 64.

SALJE, E. (1985). Thermodynamics of sodium feldspar I: order parameter treatment and strain induced coupling efffects. *Phys. Chem. Min.* **12**, 93.

SALJE, E. (1986). Raman spectroscopic investigaton of the order parameter behaviour of hypersolvus alkali feldspar: displacive phase transition and evidence for Na-K site ordering. *Phys. Chem. Min.* **13**, 340.

SALJE, E. (1987a). Thermodynamics of plagioclase I: theory of the $I\bar{1}$-$P\bar{1}$ phase transition in anorthite and Ca-rich plagioclases. *Phys. Chem. Min.* **14**, 181.

SALJE, E. (1987b). Thermodynamics of plagioclase II: temperature evolution of the spontaneous strain at the $I1$-$P\bar{1}$ phase transition in anorthite. *Phys. Chem. Min.* **14**, 181.

SALJE, E. (1988a). Kinetic rate law as derived form order parameter theory I: theoretical concepts. *Phys. Chem. Min.* **15**, 336.

SALJE, E. (1988b). Structural phase transitions and specific heat anomalies. In *Physical properties and thermodynamic behaviour of minerals* (Ed. E. Salje). Dordrecht: Reidel.

SALJE, E. & HOPPMAN, G. (1976). Direct observation of ferroelasticity in $Pb_3(PO_4)_2$-$Pb_3(VO_4)_2$. *Mat. Res. Bull.* **11**, 1545.

SALJE, E. & DEVARAJAN, V. (1981). Potts model and phase transitions in lead phosphate $Pb_3(PO_4)_2$. *J. Phys. C. Lett.*. **14**, L10.

SALJE, E. & DEVARAJAN, V. (1986). Phase transitions in systems with strain-induced coupling between two order parameters. *Phase Transitions* **6**, 235.

SALJE, E. & WERNECKE, C. (1982). The phase equilibrium between sillimanite and andalusite as determined from lattice vibrations. *Contrib. Mineral. Petrol.* **79**, 56.

SALJE, E. & WRUCK, B. (1983). Specific heat measurements and critical exponents of the ferroelastic phse transition in $Pb_3(PO_4)_2$ and $Pb_3(P_xAs_{1-x}O_4)_2$. *Phys. Rev.* **28**, 6510.

SALJE, E. & WRUCK, B. (1988). Kinetic rate laws as derived from order parameter theory II: interpretation of experimental data by Laplace-transformation, the relaxation spectrum and kinetic gradient coupling between two order parameters. *Phys. Chem. Min.* **16**, 140.

SALJE, E. & KROLL, H. (1990). Kinetic rate laws as derived from order parameter theory III: Al,Si ordering in Sanidine. *Phys. Chem. Min.* (submitted).

SALJE, E., KUSCHOLKE, B. & WRUCK, B. (1985a). Domain wall formation in minerals : I Theory of twin boundary shapes in Na-feldspar. *Phys. Chem. Min.* **12**, 132.

SALJE, E., KUSCHOLKE, B., WRUCK, B. & KROLL, H. (1985b). Thermodynamics of sodium feldspar II: experimental results and numerical calculations. *Phys. Chem. Min.* **12**, 99.

SALJE, E., BISMAYER, U. & JANSEN, M. (1987a). Temperature evolution of the ferroelastic order parameter of As_2O_5 as determined from optical birefringence. *J. Phys. C* **20**, 3613.

SALJE, E., PALOSZ, B. & WRUCK, B. (1987b). In-situ observation of the polytypic phase transition 2H-12R in PbI_2: investigation of the thermodyanmic structural and dielectric properties. *J. Phys. C.* **20**, 4077.

SAMARA, G.A. (1978). Comment on tricritical point in SbSi - note. *Phys. Rev. B.* **17(7)**, 3020.

SANDERCOCK, J.R., PALMER, S.B., ELLIOTT, R.J., HAYES, W., SMITH, S.R.P. & YOUNG, A.P. (1972). Brillouin scattering, ultrasonic and theoretical studies of acoustic anomalies in crystals showing Jahn-Teller phase transitions. *J. Phys. C* **5(21)**, 3126.

SAPRIEL, J. (1975). Domain wall orientations in ferroelastics. *Phys. Rev.* **B12(11)**, 5128.

SAUNDERS, G.A., COMINS, J.D., MACDONALD, J.E. & SAUNDERS, E.A. (1986). Thermodynamics of a ferroelastic phase transition. *Phys. Rev.* **B34**, 2064.

SCHMAHL, W.W. & SALJE, E. (1989). X-ray diffraction study of the orientational order/disorder transition in $NaNO_3$: evidence for order parameter coupling. *Phys. Chem. Min.* **16**, 790.

SCHMAHL, W.W. & REDFERN, S.A.T. (1988).An X-ray study of coupling between acoustic and optic modes at the ferroelastic phase transition in As_2O_5. *J. Phys. C.* **21**, 3719.

SCHMAHL, W.W., PUTNIS, A., SALJE, E., FREEMAN, P., GRAEME BARBER, A., JONES, R., SINGH, K.K., BLUNT, J., EDWARDS, P.P., LORAM, J. &

MISZA, K. (1989). Twin formation and structural modulations in orthorhombic and tetragonal YBa$_2$(Cu$_{1-x}$Co$_x$)O$_{7-\delta}$. *Phil. Mag. Lett.* **60**, 6241.

SCHMID, H. BURKHARD, E. ,SUN, B.N. & RIVERA, J.P. et al. (1989) Uniaxial stress-induced ferroelastic detwinning of YBa$_2$Cu$_3$O$_{7-\gamma}$. *Physica C* **157**, 555.

SCHMIDT, V.H., WESTERN, A.B., BAKER, A.G. & BACON, C.R. (1978). Tricritical point and tricritical exponent - data in KH$_2$PO$_4$. *Ferroelectr.* **20(3-4)**, 169.

SCOTT, J.F., HABBAL, F. & HIDAKA, M. (1982). Phase transitions in BaMnF$_4$ specific heat. *Phys. Rev.* **B25**, 1805.

SETO, H., NODA, Y. & YAMADA, (1989). Precursor phenomena at martensitic phase transition in Fe-Pd alloy. II: Huang scattering and embryonic fluctuations. *Tech. Rep. of ISSP Ser A.* **No.2119**, 11.

SHANG, H.T. & SALOMON, M.B. (1980). Tricritical scaling and logarithmic corrections for metamagnet FeCl$_2$. *Phys. Rev. B.* **22**, 4401.

SLONCZEWSKI, J.C. & THOMAS, H. (1970). Interaction of elastic strain with structural transition of strontium-titanate. *Phys. Rev.* **B1(9)**, 3599.

SMITH, J.V. (1974). *Feldspar minerals.* Berlin: Springer-Verlag.

SPEER, D. & SALJE, E. (1986). Phase transitions in langbeinites I: crystal chemistry and structures of K-double sulphates of the Langbeinite type M$_2^{++}$K$_2$(SO$_4$)$_3$, M^{++}=Mg,Ni,Co,Zn,Ca. *Phys. Chem. Min.* **13**, 17.

STOKES, H.T. & HATCH, D.M. (1988). *Isotropy subgroups of the 230 crystallographic space groups.* Singapore: World Scientific.

STRUKOV, B.A., TARASKIN, J.A., MINAEVA, K.A. & FEDORIKH, V.A. (1980). Critical phenomena in perfect and imperfect TGS crystals. *Ferroelectr.* **25**, 399.

SUZUKI, T. & WUTTIG, M. (1975). Martensitic transformation and lattice instability. *B. Am. Phys.* **20(3)**, 349.

SUZUKI, T. & ISHIBASHI, Y. (1987). Phenomenological considerations of the electric field induced transitions in improper ferroelectrics and ferroelastics. III. Application to Gd$_2$(MoO$_4$)$_3$. *J. Phys. Soc. Jap.* **56(2)**, 596.

TANG, S., MAHANTI, S.D. & KALIA, R.K. (1986). Ferroelastic phase transition in two-dimensional molecular solids. *Phys. Rev. Letters* **56**, 484.

TENTRUP, T. & SIEMS, R. (1986). Structure and free energy of domain walls in ANNNI systems. *J. Phys. C.* **19**, 3443.

THOEN, J. (1978). Assymetry in homogeneous phase specific-heats at constant volume on coexistence curve near critical point of simple fluids. *Phys. Chem. Letters* **8(2)**, 99.

TOLEDANO, J.C. (1979). Phenomenological model for the structural transition in Benzil. *Phys. Rev.* **B20**, 1147.

TOLEDANO, P. & TOLEDANO, J.C. (1976). Order parameter symmetries for improper ferroelectric non-ferroelastic transitions. *Phys. Rev.* **B14**, 3097.

TOLEDANO, P. & TOLEDANO, J.C. (1977). Order parameter symmetries for phase transitions of nonmagnetic secondary and higher order ferroics. *Phys. Rev.* **B16**, 386.

TOLEDANO, J.C. & TOLEDANO, P. (1980). Order parameter symmetries and free energy expansions for purely ferroelastic transitions. *Phys. Rev.* **B21(3)**, 1139.

TOLEDANO, P. & TOLEDANO, J.C. (1982). Non-ferroic phase transitions. *Phys. Rev.* **B25(3)**, 1946.

TOLEDANO, J.C. & TOLEDANO, P. (1988). *The Landau theory of phase transitions*. World Scientific.

TORRES, J., ROUCAN, C. & AYROLES, R. (1982a). Investigation of the interactions between ferroelastic domain walls and of the structural transition in lead phosphate observed by electron microscopy. I: Experimental results. *Phys. Stat Sol.* **A70.**, 659. '

TORRES, J., ROUCAN, C. & AYROLES, R. (1982b). Investigation of the interactions between ferroelastic domain walls and of the structural transition in lead phosphate observed by electron microscopy. III: Interpretation of the interactions between ferroelastic domain walls. *Phys. Stat. Sol.* **A70**, 193.

WADA, M., SAWADA, A., ISHIBASHI, Y. & TAKAGI, Y. (1977). Raman-scattering spectra of K_2SeO_4. *J. Phys. Soc. Jap.* **42**, 1229.

WADA, M., UWE, H., SAWADA, A., ISHIBASHI, Y., TAKAGI, Y. & SAKUDO, T. (1977). Lower frequency soft mode in ferroelectric phase of K_2SeO_4. *J. Phys. Soc. Jap.* **43**, 544.

WADWAHAN, V.K. (1978). Ferroelastic effect in orthoboric acid. *Mater. Res. B* **13(1)**: 1-8.

WADWAHAN, V.K. (1982). Ferroelasticity and related properties of crystals. *Phase Transitions* **3**, 3.

WENYUAN, S., HUIMIN, S., YENING, W. & BAOSHENG, L. (1985). Internal friction associated with domain walls and ferroelastic transition in LNPP. *J. Physique* **46** (suppl.12), C10.

WINTERFELD, T.V. & SCHAAK, G. (1977). Phonon contributions to anomalous specific heat of triglycine sulfate near T_c. *Phys. Stat. Sol.* **B80(2)**, 559.

WOOSTER, W.A. & WOOSTER, N. (1946). Control of electrical twinning in quartz - letter. *Nature* **157(3987)**, 405.

WRUCK, B. (1986). Thesis, University of Hannover.

WUERZ, U. & GRUBIC, M. (1980). An adiabatic calorimeter of the scanning ratio type. *J. Phys.* **E13**, 525.

YAMADA, Y., NODA, Y., TAKIMOTO, M. & FURUKAWA, K. (1985). Modulated lattice relaxation and incommensurability of lattice waves in a beta-based premartensite phase. *J. Phys. Soc Jap.* **54**, 2940.

YAMAMOTO, N., YAGI, K. & HONJO, G. (1977a). Electron microscopic studies of ferroelectric and ferroelastic $Gd_2(MoO_4)_3$. I: General features of ferroelectric domain wall, antiphase boundary and crystal defects. *Phys. Stat. Sol.* **A41**, 523.

YAMAMOTO, N., YAGI, K. & HONJO, G. (1977b). Electron microscopic studies of ferroelectric and ferroelastic $Gd_2(MoO_4)_3$. II: Interactions between ferroelectric domain walls. *Phys. Stat. Sol.* **A42**, 257.

YOSHIHARA, A., YOSHIZAWA, M., YASUDA, H. & FUJIMURA, T. (1985). On the 83K ferroelastic phase transition in Benzil. *Jap. J. Appl. Phys. Suppl.* **24-2**, 367.

ZOO, J.D. & LIU, S.T. (1976). Use of effective field theory to predict relationships among ferroelectric parameters. *Ferroelectr.* **11(1-2)**, 371.

APPENDIX

An atomistic model for the Landau potential and the origin of the saturation effect

Let us consider the lattice instability along a specific direction in the reciprocal lattice. All lattice planes perpendicular to this direction are projected on this one-dimensional subspace, each plane marking a point in a chain. Each point is then characterized by an effective local potential and collective interactions between the points. The points are numbered l and the total potential can be written in terms of the local and normal coordinates Q_l and conjugate momenta P_l. The equivalent Hamiltonian is

$$H = \sum_l \left[\frac{1}{2M} P_l^2 + \tfrac{1}{2} M\Omega_0^2 Q_l^2 + \tfrac{1}{4} u Q_l^4 + \dots \right] - \tfrac{1}{2} \sum_{l \neq l'} v_{ll'} Q_l Q_{l'}$$

Here M is the effective mass, Ω_0 is the local frequency for small oscillations in the absence of interactions, u gives the strength of the anharmonicity (which is restricted to fourth order in this case) and $v_{ll'}$ are interaction constants.

The type of transition is now determined by the wavevector q_o at which the Fourier transform of the interaction assumes its maximum value. The mean-field approximation of this Hamiltonian has been discussed in detail by Salje, Wruck & Thomas (1991) and the author will follow their arguments closely. The temperature dependence of the order parameter results from the dissipation-fluctuation theorem which links the variance of the order parameter $\sigma = \langle (Q_l - \langle Q_l \rangle)^2 \rangle$ with the excitation frequency Ω via

$$\sigma = (\hbar/2M\Omega)\coth(\hbar\Omega/2k_B T)$$

leading to expectation values

$$\langle P_l^2 \rangle = (M\Omega)^2 \sigma$$
$$\langle Q_l^2 \rangle = Q^2 + \sigma$$
$$\langle Q_l^4 \rangle = Q^4 + 6Q^2\sigma + 3\sigma^2$$

The Gibbs free energy becomes

$$G = N\{ \tfrac{1}{2}(M\Omega_0^2 - v + 3u\sigma)Q^2 + \tfrac{1}{4}uQ^4 + \tfrac{1}{2}M\Omega_0^2\sigma + \tfrac{3}{4}u\sigma^2$$
$$- \tfrac{1}{4}\hbar\Omega\coth\left(\frac{\hbar\Omega}{2k_B T}\right) + k_B T \ln\left[2\sinh\left(\frac{\hbar\Omega}{2k_B T}\right)\right] \}$$

Minimizing with respect to Q and Ω yields the mean-field equations:

$$(M\Omega_0^2 - v + 3u\sigma + uQ^2)Q = 0$$

$$M\Omega^2 = M\Omega_0^2 + 3u(\sigma + Q^2)$$

The variance at the critical point is found by letting Q go to zero:

$$\sigma_c = \frac{(v - M\Omega_0^2)}{3u}$$

and for the order parameter in the low-symmetry phase one obtains

$$Q^2 = 3(\sigma_c - \sigma).$$

The self-consistency condition for σ is:

$$\sigma = \sigma_c \left(\frac{\eta\Omega_0}{\Omega}\right) \coth\left(\frac{\eta\Omega}{\Omega_0}\right)$$

$$\Omega^2 = \Omega_0^2 \left[1 + \Delta\left(3 - \frac{2\sigma}{\sigma_c}\right)\right]$$

where the normalised temperature x has been introduced as

$$x = \frac{k_B T}{M\Omega_0^2 \sigma_c}$$

The dimensionless parameter

$$\Delta = \frac{v}{M\Omega_0^2} - 1$$

is proportional to the distance from the displacive limit with $v = M\Omega_0^2$ and

$$\eta = \frac{\hbar}{2M\Omega_0 \sigma_c}$$

is a measure of the quantum influence at low temperatures. The quantum-mechanical fluctuation enhancements have an important influence on the system because the zero-point fluctuations with variance σ_s reduce the order parameter Q_s at zero temperature, as compared to its classical value Q_0 $(= 3\sigma_c^2)$ by

$$\frac{Q_s^2}{Q_0^2} = 1 - \frac{\sigma_s}{\sigma_c}$$

The value of σ_s is determined by the solution of the equation

$$\left(\frac{\sigma_s}{\sigma_c}\right)\left[1 + \Delta\left(3 - \frac{2\sigma_s}{\sigma_c}\right)\right]^{\frac{1}{2}} = \eta$$

which follows from the self-consistency equation in the limit T = 0 K. Simultaneously, the quantum fluctuations reduce T_c by

$$\frac{T_c}{T_c^{class}} = \frac{\eta/(1 + \Delta)^{\frac{1}{2}}}{arctanh[\eta/(1 + \Delta)^{\frac{1}{2}}]}.$$

So far the phase transition has been assumed to be second order and higher order anharmonicities of Q_l were ignored. Salje, Wruck & Thomas (1991) have shown that similar arguments hold in the displacive limit for a Hamiltonian including a sixth-order term. If the sixth-order term is present but the fourth-order term vanishes, one finds a classical tricritical phase transition.

The classical Landau potentials follow now as the high temperature approximations of the Gibbs free energies developed so far. Salje, Wruck & Thomas (1991) have shown that in the displacive limit (for which Landau theory is applicable) the Gibbs free energy becomes

$$G(T,Q) = G_0(T) + \tfrac{3}{2}[\sigma(T) - \sigma_c]uQ^2 + \tfrac{1}{4}uQ^4 + \tfrac{1}{6}cQ^6$$

where the mean-square fluctuations are given by

$$\sigma(T) = \frac{k_B\Theta_s}{M\Omega_0^2} coth\left(\frac{\Theta_s}{T}\right)$$

in terms of the characteristic temperature $\Theta_s = 2T_s$ defined by

$$k_B\Theta_s = \tfrac{1}{2}\hbar\Omega_0.$$

With the parameters $A = 3uk_B/m\Omega_0^2$, $B = u$ and the critical temperature T_c defined by

$$\Theta_s coth\left(\frac{\Theta_s}{T_c}\right) = \frac{M\Omega_0^2Q_0^2}{3k_B}$$

the Landau potential becomes in its general form

$$G(Q,T) = G_0(T) + \tfrac{1}{2}A\Theta_s[coth\frac{\Theta_s}{T} - coth\frac{\Theta_s}{T_c}]Q^2 + \tfrac{1}{4}BQ^4 + \ldots$$

In the high-temperature approximation $(T \gg T_s)$ this equation yields

$$G(T,Q) = L(T,Q) = \tfrac{1}{2}A(T - T_c)Q^2 + \tfrac{1}{4}BQ^4 + \tfrac{1}{6}cQ^6$$

Table 1a *Group subgroup relationships for equitranslational phase transitions*

group	subgroup		active rep.	spontaneous strain proper	improper	elastic domains	basis functions	
triclinic								
$\bar{1}$	1	c	A_u			1	x, y, z	
monoclinic								
2	1	c	B	e_4, e_6		2	z, x, $\underline{z}$, $\underline{x}$	[yz],[xy]
m	1	c	A"	e_4, e_6		2	y, $\underline{z}$, $\underline{x}$	[yz],[xy]
2/m	2	c	A_u			1	y	
2/m	$\bar{1}$	c	B_g	e_4, e_6		2	z, x,	[yz],[xy]
2/m	m	c	B_u			1	y, z	
o'rhombic								
222	2	c	B_1	e_6		2	z, $\underline{z}$,	[xy]
222	2	c	B_2	e_5		2	y, $\underline{y}$	[xz]
222	2	c	B_3	e_4		2	x, $\underline{x}$	[yz]
mm2	2	c	A_2	e_6		2	z	[xy]
mm2	m	c	B_1	e_5		2	x, $\underline{y}$,	[xz]
mm2	m	c	B_2	e_4		2	y, $\underline{x}$,	[yz]
mmm	222	c	A_u			1	x[yz],y[xz],z[yx]	
mmm	2/m	c	B_{1g}	e_6		2	$\underline{z}$,	[xy]
mmm	2/m	c	B_{2g}	e_5		2	$\underline{y}$,	[xz]
mmm	2/m	c	B_{3g}	e_4		2	$\underline{x}$,	[yz]
mmm	mm2	c	B_{1u}			1	z	
mmm	mm2	c	B_{2u}			1	y	
mmm	mm2	c	B_{3u}			1	x	
tetragonal								
4	2	c	B	$e_6,$ $e_1 = -e_2$		2	[xx-yy]	[xy]
4	1	c	E	e_4, e_5	$e_6,$ $e_1 = -e_2$	4	x, $\underline{x}$, y, $\underline{y}$,	[xz] [yz]

Table 1a   *Cont'd*

group	subgroup		active rep.	spontaneous strain proper	spontaneous strain improper	elastic domains	basis functions	
$\bar{4}$	2	c	B	$e_6,$ $e_1 = -e_2$		2	z,[xx-yy,]	[xy]
$\bar{4}$	1	c	E	e_4, e_5	$e_6,$ $e_1 - -e_2$	4	x, y, y, x,	[yz] [xz]
4/m	4	c	A_u			1	z	
4/m	2/m	c	B_g	$e_6,$ $e_1 = -e_2$		2	[xx-yy],	[xy]
4/m	$\bar{4}$	c	B_u			1	x[yz]+y[xz], z[xy], y[yz]-x[xz] z[xx-yy]	
4/m	$\bar{1}$	c	E_g	e_4, e_5	$e_6,$ $e_1 - -e_2$	4	x, y,	[xz] [yz]
4/m	m	c	E_u		$e_6,$ $e_1 - -e_2$	2	x, y	
422	4	c	A_2			1	z, z	
422	222	c	B_1	$e_1 = -e_2$		2	[xx-yy]	
422	222	c	B_2	e_6		2		[xy]
422	2	c	E	e_4	$e_1 = -e_2$	4 ⎫		
	2	c	E	$e_4 = -e_5$	e_6	4 ⎬	x, x	[yz]
	1	c	E	$e_4,$ e_5	$e_6,$ $e_1 = -e_2$	8 ⎭	y, y	-[xz]
4mm	4	c	A_2			1	z, x[yz]-y[xz]	
4mm	mm2	c	B_1	$e_1 = -e_2$		2	[xx-yy]	
4mm	mm2	c	B_2	e_6		2		[xy]
4mm	m	c	E	e_4	$e_1 = -e_2$	4 ⎫		
	m	c	E	$e_4 = -e_5$	e_6	4 ⎬	x, y,	[xz]
	1		E	$e_4,$ e_5	$e_6,$ $e_1 = -e_2$	8 ⎭	y, -x	[yz]

Table 1a *Cont'd*

group	subgroup		active rep.	spontaneous strain proper	improper	elastic domains	basis functions	
$\overline{4}2m$	$\overline{4}$	c	A_2			1	$z,y[yz] - x[xz],z[xx-yy]$	
$\overline{4}2m$	222	c	B_1	$e_1 = -e_2$		2	$[xx-yy]$	
$\overline{4}2m$	mm2	c	B_2	e_6		2	$z,$	$[xy]$
$\overline{4}2m$	m	c	E	$e_4 = -e_5$	e_6	4⎫		
	2	c	E	e_4	$e_1 = -e_2$	4⎬	$x,\underline{x}$	$[yz]$
	1		E	$e_4,$	$e_6,$	8⎭	$y,\underline{y},$	$[xz]$
				e_5	$e_1 = -e_2$			
4/mmm	422	c	A_{1u}			1	$x[yz] - y[xz]$	
4/mmm	4/m	c	A_{2g}			1	$z, [xx-yy][xy]$	
4/mmm	4mm	c	A_{2u}			1	z	
4/mmm	mmm	c	B_{1g}	$e_1 = -e_2$		2	$[xx-yy]$	
4/mmm	mmm	c	B_{2g}	e_6		2		$[xy]$
4/mmm	$\overline{4}2m$	c	B_{1u}			1	$x[yz] + y[xz],z[xy]$	
4/mmm	$\overline{4}2m$	c	B_{2u}			1	$x[xz] - y[yz],z[xx-yy]$	
4/mmm	2/m	c	E_g	$e_4 = -e_5$	e_6	4⎫		
	2/m	c		e_4	$e_1 = -e_2$	4⎬	$\underline{x},$	$[yz]$
	$\overline{1}$			$e_4,$	$e_6,$	8⎭	$\underline{y},$	$-[xz]$
				e_5	$e_1 = -e_2$			
4/mmm	2mm	c	E_u		e_6	2⎫		
	2mm	c			$e_1 = -e_2$	2⎬	x	
	m				$e_6,$	4⎭	y	
					$e_1 = -e_2$			

trigonal

3	1		E	$e_k, k>3,$		3	$x,\underline{x},[xx-yy],[xz]\,\underline{y},y, -2[xy],[yz]$	
				$e_1 = -e_2$				
$\overline{3}$	3	c	A_u			1	z	
$\overline{3}$	$\overline{1}$		E_g	$e_k, k>3,$		3	$\underline{x},[xx-yy],$	$[xz]$
				$e_1 = -e_2$			$\underline{y}, -z[xy]$	$[yz]$
$\overline{3}$	1	c	E_u		$e_k, k>3$	3	x, y	
					$e_1 = -e_2$			

Table 1a   *Cont'd*

group	subgroup		active rep.	spontaneous strain proper	spontaneous strain improper	elastic domains	basis functions
32	3	c	A_2			1	z, z̲
32	2		E	e_4		3⎫	
				$e_1 = -e_2$			x,x̲,[xx-yy], [yz]
	1		E	$e_k, k>3,$		6⎰	y,y̲, -2[xy],-[xz]
				$e_1 = -e_2$			
3m	3	c	A_2			1	z̲,x[xx-yy]-2y[xy],x[yz]-y[xz]
3m	m		E	$e_4,$		3⎫	
				$e_1 = -e_2$			x,y̲, 2[xy],[xz]
	1		E	$e_k, k>3,$		6⎰	y,-x̲,[xx-yy], [yz]
				$e_1 = -e_2$			
3̄m	32	c	A_{1u}			1	x[yz] - y[xz],x[xx-yy] - 2y[xy]
3̄m	3̄	c	A_{2g}			1	z̲, [xx-yy][xz] - 2[xy][yz]
3̄m	3m	c	A_{2u}			1	z
3̄m	2/m		E_g	$e_4,$		3⎫	
				$e_1 = -e_2$			x,[xx-yy], [yz]
	1̄		E_g	$e_k, k>3,$		6⎰	y̲, -2[xy],-[xz]
				$e_1 = -e_2$			
3̄m	m	c	E_u		$e_4,e_1 = -e_2$	3⎫	
	2	c	E_u		$e_4,e_1 = -e_2$		x
	1		E_u		$e_k, k>3,$	6⎰	y
					$e_1 = -e_2$		

exagonal

group	subgroup		active rep.	spontaneous strain proper	spontaneous strain improper	elastic domains	basis functions
6	3	c	B			1	x[xx-yy]-2y[xy], 2x[xy]+y[xx-yy]
6	2	c	E_1	$e_6,$		3⎫	2[xy]
				$e_1 = -e_2$		⎰	[xx-yy]
6	1	c	E_2	$e_4,$	$e_6,$	6⎫	x,x̲, [xz]
				e_5	$e_1 = -e_2$	⎰	y,y̲, [yz]

Table 1a   *Cont'd*

group	subgroup	active rep.	spontaneous strain proper	spontaneous strain improper	elastic domains	basis functions	
$\bar{6}$	3 c	A"			1	z	
$\bar{6}$	m c	E'	$e_6,$ $e_1 = -e_2$		3⎫ ⎬	x, y,[xx-yy]	2[xy]
$\bar{6}$	1 c	E"	$e_4,$ e_5	$e_6,$ $e_1 = -e_2$	6⎫ ⎬	x, y,	[xz] [yz]
6/m	6 c	A_u			1	z	
6/m	$\bar{3}$ c	B_g			1	[xx-yy][yz] + 2[xy][xz] [xx-yy][xz] - 2[xy][yz]	
6/m	$\bar{6}$ c	B_u			1	x[xx-yy]-2y[xy],2x[xy]+y[xx-yy]	
6/m	2/m c	E_{1g}	e_6 $e_1 = -e_2$		3⎫ ⎬	[xx-yy]	2[xy]
6/m	2 c	E_{1u}		$e_6,$ $e_1 = -e_2$	3⎫ ⎬	z[xx-yy],-x[xz]+y[yz] -z[xy],x[yz]+y[xz]	
6/m	$\bar{1}$ c	E_{2g}	$e_4,$ e_5	$e_6,$ $e_1 = -e_2$	6⎫ ⎬	x, y,	[xz] [yz]
6/m	m c	E_{2u}		$e_6,$ $e_1 = -e_2$	3⎫ ⎬	x y	
622	6 c	A_2			1	z, z	
622	32 c	B_1			1	x[xx-yy] - 2y[xy]	
622	32 c	B_2			1	2x[xy] + y[xx-yy]	
622	2	E_1	e_5	$e_1 = -e_2$	6		
	2	E_1	e_4	$e_1 = -e_2$	6	x,x, y,y,	[yz] -[xz]
	1		$e_4,$ e_5	$e_6,$ $e_1 = -e_2$	12		
622	222	E_2	$e_1 = -e_2$		3⎫ ⎬	[xx-yy]	2[xy]
	2	E_2	$e_6,$ $e_1 = -e_2$		6⎭		

Table 1a   *Cont'd*

group	subgroup		active rep.	spontaneous strain proper	spontaneous strain improper	elastic domains	basis functions	
6mm	6	c	A_2			1	$z, x[yz] - y[xz]$	
6mm	3m	c	B_1			1	$x[xx-yy] - 2y[xy]$	
6mm	3m	c	B_2			1	$2x[xy] + y[xx-yy]$	
6mm	m	c	E_1	e_5	$e_1 = -e_2$	6 ⎫		
	m	c	E_1	e_4	$e_1 = -e_2$	6 ⎬	x,y	[xz]
	1		E_1	$e_4,$ e_5	$e_6,$ $e_1 = -e_2$	12 ⎭	$y,-x,$	[yz]
6mm	mm2		E_2	$e_1 = -e_2$		3 ⎫	[xx-yy]	
	2		E_2	$e_6,$ $e_1 = -e_2$		6 ⎬⎭		$-2[xy]$

$\bar{6}m2$	$\bar{6}$	c	A'_2			1	$z, x[xx-yy] - 2y[xy]$	
$\bar{6}m2$	32	c	A''_1			1	$x[yz] - y[xz]$	
$\bar{6}m2$	3m	c	A''_2			1	z	
$\bar{6}m2$	m2m		E'	$e_1 = -e_2$		3 ⎫	$x,$	[2xy]
	m		E'	$e_6,$ $e_1 = -e_2$		6 ⎬⎭	$y,[xx-yy]$	
$\bar{6}m2$	m	c	E''	e_4	$e_1 = -e_2$	6 ⎫		[xz]
	2	c	E''	e_5	$e_1 = -e_2$	6 ⎬	y	[xz]
	1		E''	$e_4,$ e_5	$e_6,$ $e_1 = -e_2$	12 ⎭	$-x,$	[yz]

6/mmm	622	c	A_{1u}			1	$x[yz] - y[xz]$	
6/mmm	6/m	c	A_{2g}			1	$z,\{[xz][yz]\}, \{[xx-yy][xy]\}$	
6/mmm	6mm	c	A_{2u}			1	z	
6/mmm	$\bar{3}m$	c	B_{1g}			1	$[xx-yy][yz] + 2[xy][xz]$	
6/mmm	$\bar{3}m$	c	B_{2g}			1	$[xx-yy][xz] - 2[xy][yz]$	
6/mmm	$\bar{6}2m$	c	B_{1u}			1	$x[xx-yy] - 2y[xy]$	
6/mmm	$\bar{6}m2$	c	B_{2u}			1	$2x[xy] + y[xx-yy]$	
6/mmm	$\bar{1}$		E_{1g}	$e_4,$ e_5	$e_6,$ $e_1 = -e_2$	12 ⎫	$x,$	[yz]
	2/m	c	E_{1g}	e_4	$e_1 = -e_2$	6 ⎬	$y,$	-[xz]
	2/m	c	E_{1g}	e_5	$e_1 = -e_2$	6 ⎭		

Table 1a   *Cont'd*

group	subgroup	active rep.	spontaneous strain proper	spontaneous strain improper	elastic domains	basis functions
6/mmm	m2m c	E_{1u}		$e_1 = -e_2$	3 ⎫	x
	2mm c	E_{1u}		$e_1 = -e_2$	3 ⎬	y
	m	E_{1u}		$e_6,$	6 ⎭	
				$e_1 = -e_2$		
6/mmm	mmm	E_{2g}	$e_1 = -e_2$		3 ⎫	2[xy]
	2/m	E_{2g}	e_6		6 ⎬	[xx-yy]
			$e_1 = -e_2$		⎭	
6/mmm	mm2 c	E_{2u}		$e_1 = -e_2$	3 ⎫	z[xx-yy],x[xz]-y[yz]
	222 c	E_{2u}		$e_1 = -e_2$	3 ⎬	-2z[xy],-x[yz]+y[xz]
	2	E_{2u}		$e_6,$	6 ⎭	
				$e_1 = -e_2$		

cubic

group	subgroup	active rep.	spontaneous strain proper	spontaneous strain improper	elastic domains	basis functions
23	222	E			3 ⎫	√3[xx-yy]
			$e_0,$		⎬	[2zz-xx-yy]
			e_t		⎭	
23	3	T	$e_4 = e_5$		4 ⎫	x,$\underline{x}$ [yz]
			$= e_6$		⎪	y,$\underline{y}$, [xz]
	2	T	e_6		6 ⎬	z,$\underline{z}$, [xy]
				$e_0,$	⎪	
				e_t	⎪	
	1	T	e_4		12 ⎪	
			e_5	$e_0,$	⎭	
			e_6	e_t		

Table 1a　*Cont'd*

group	subgroup	active rep.	spontaneous strain proper	spontaneous strain improper	elastic domains	basis functions
m3	23 c	A_u			1	$x[yz] + y[xz] + z[xy]$
m3	mmm	E_g			3	
			$e_o,$ e_t			
				e_o e_t	3 ⎱	$\sqrt{3}[xx\text{-}yy]$ $[2zz\text{-}xx\text{-}yy]$
m3	222 c	E_u			3 ⎱	$\sqrt{3}(x[yz]\text{-}y[zx])$ $2z[xy]\text{-}x[yz]\text{-}y[xz]$
n3	$\bar{3}$	T_g	$e_4 = e_5$ e_6		4 ⎱	
	$2/m$	T_g	e_6		6	$\underline{x},$　　[yz]
				$e_o,$ e_t		$\underline{y},$　　[xz]
	$\bar{1}$	T_g	$e_4,$ $e_5,$ e_6	$e_o,$ e_t	12	$\underline{z}$　　　[xy]
m3	3 c	T_u			4 ⎱	
	mm2 c	T_u		$e_o,$ e_t $e_1 + e_2$	3	
	m	T_u		$e_6,$	6	x
				$e_o,$ e_t		y z
	1	T_u		$e_1, e_2, e_4,$ $e_5, e_6,$ e_o, e_t	12	

Table 1a   *Cont'd*

group	subgroup	active rep.	spontaneous strain proper	spontaneous strain improper	elastic domains	basis functions
432	23 c	A_2			1	$x[yz] + y[zx] + z[xy]$
432	422	E	e_t		3	$\sqrt{3}\,[xx-yy]$
	222	E	e_o, e_t		6	$[2zz-xx-yy]$
432	3	T_1		$e_4=e_5=e_6$	4	
	4	T_1		e_t	3	
	2	T_1			12	$x, \underline{x}$
				e_t		$y, \underline{y}$
				$e_4 = -e_5,$ e_6		$z, \underline{z}$
	1	T_1			24	
				$e_o,$ e_t e_4, e_5, e_6		
432	32	T_2	$e_4=e_5=e_6$		4	
	222	T_2	e_6		6	
				e_t		$[yz]$
	2	T_2	$e_4=-e_5$		12	$[xz]$
			e_6	e_t		$[xy]$
	1	T_2	e_4,e_5,e_6		24	
				$e_t,$ e_o		
$\bar{4}3m$	23 c	A_2			1	$\{xxyy\} + \{yyzz\} + \{zzxx\}$
$\bar{4}3m$	$\bar{4}2m$	E	e_t		3	$\sqrt{3}[xx-yy]$
	222	E			6	$[2zz-xx-yy]$
			$e_t,$ e_o			

Table 1a   *Cont'd*

group	subgroup	active rep.	spontaneous strain proper	spontaneous strain improper	elastic domains	basis functions
$\bar{4}3m$	3 c	T_1		$e_4=e_5=e_6$	4	
	$\bar{4}$ c	T_1			3	x,x[yy-zz],z[zx]-y[yx]
				e_t		y,y[zz-xx],x[xy]-z[zy]
	m	T_1		e_t	12	z,z[xx-yy],y[yz]-x[xz]
				$e_4 = e_5,$ e_6		
	1	T_1			24	
				$e_0,$ e_t e_4, e_5, e_6		
$\bar{4}3m$	3m	T_2	$e_4=e_5=e_6$		4	
	mm2	T_2	e_6		6	x, [yz]
				e_t		y, [xz]
	m	T_2	$e_4 = -e_5,$ e_6	e_t	12	z, [xy]
	1	T_2	e_4,e_5,e_6		24	
				$e_0,$ e_t		
m3m	432 c	A_{1u}			1	x{yz} + y{zx} + z{xy}
m3m	m3 c	A_{2g}			1	{xxyy} + {yyzz} + {zzxx}
m3m	$\bar{4}3m$ c	A_{2u}			1	x[yz] + y[zx] + x[xy]
m3m	4/mmm	E_g			3	$\sqrt{3}$[xx-yy]
				e_t		
	mmm	E_g			6	[2zz-xx-yy]
				$e_0,$ e_t		

Table 1a *Cont'd*

group	subgroup		active rep.	spontaneous strain proper	improper	elastic domains	basis functions
m3m	$\bar{4}2m$	c	E_u		e_t	3	
	422	c	E_u		e_t	3	$x[yz]+y[zx]-2z[xy]$
							$\sqrt{3}(x[yx]-y[zx])$
	222		E_u		$e_o,$	6	
					e_t		
m3m	$\bar{3}$	c	T_{1g}		$e_4=e_5=e_6$	4	
	4/m	c	T_{1g}		e_t	3	
	2/m		T_{1g}		e_t	12	$\underline{x},[yy-zz]\ [yz]$
					$e_4=-e_5,$		$\underline{y},[zz-xx]\ [zx]$
					e_6		
	$\bar{1}$		T_{1g}		$e_o,$		$\underline{z},[xx-yy]\ [xy]$
					e_t		
					e_4,e_5,e_6	24	

Table 1a *Cont'd*

group	subgroup	active rep.	spontaneous strain proper	spontaneous strain improper	elastic domains	basis functions
m3m	3m c	T_{1u}		$e_4=e_5=e_6$	4	
	4mm c	T_{1u}		e_t	3	
	2mm	T_{1u}		e_t	6	
	m	T_{1u}		e_6 $e_o,$ e_t	12	
	m	T_{1u}		e_6 e_t	24	x y z
	1	T_{1u}		$e_4=-e_5$ e_6 $e_o,$ e_t e_4,e_5,e_6	24	
m3m	$\bar{3}$m	T_{2g}	$e_4=e_5=e_6$		4	
	mmm	T_{2g}	e_6	e_t	6	[yz]
	2/m	T_{2g}	$e_4=-e_5,$ e_6	e_t	12	[xz]
	$\bar{1}$	T_{2g}	e_4,e_5,e_6	$e_o,$ e_t	24	[xy]

Table 1a   *Cont'd*

group	subgroup	active rep.	spontaneous strain proper	spontaneous strain improper	elastic domains	basis functions
m3m	32 c	T_{2u}		$e_4=e_5=e_6$	4	
	$\bar{4}m2$ c	T_{2u}		e_t	3	
	2mm	T_{2u}			6	
				e_t		
				e_6		
	m	T_{2u}		$e_o,$	12	$x[yy-zz],z[zx]-y[yz]$
				e_t		$y[zz-xx],x[xy]-z[zy]$
						$z[xx-yy],y[yz]-x[xz]$
				e_6	12	
	2	T_{2u}		e_t		
				$= -2e_3$		
				$e_4 = -e_5,$		
				e_6	24	
	1	T_{2u}		$e_o,$		
				e_4,e_5,e_6		

*This list includes all cases of proper ferroelastic or co-elastic phase transitions. The choice of the coordinate system follows Nye (1964) with x,y,z as a right hand system (second setting in the monoclinic system). The relationships were first published by Janovec et al. (1975). The meaning of the columns is:

Column 'group' gives the international symbol (see International Tables of Crystallography, 1980) of the point group of the high symmetry phase.

Column 'subgroup' shows the equivalent symbol of the low symmetry phase (e.g. the ferroelastic or co-elastic phase).

Column 'active rep.' contains the Mulliken symbol of the irreducible representation of the high symmetry phase which transforms as the order parameter of the phase transition between the two phases given in the first two columns. Physically irreducible representations are treated also as irreducible representations.

Column 'spontaneous strain'. Here the components of the proper (i.e. proportional Q) and improper (i.e.proportional Q^2) spontaneous strain are listed. Note the possibility for higher order coupling as discussed in the text. The components of the improper spontaneous strain which coincide with the thermal expansion strain components of the low symmetry phase are not listed because they can easily be worked out from the symmetry constraints of these phases. All components are listed in Voigt notation.

Column 'domains'. The number of ferroelastic domains is given here. This number is equal to n_G/n_s where n_s is the order of the maximal subgroup of G that is a super group of the low-symmetry point group and does not change the spontaneous strain in one domain (see Chapter 4). Each two domains have different state tensors S_{ik} (Chapter 8). If there is only one domain, no switching is possible and the crystal cannot be ferroelastic (not even 'potentially' ferroelastic).

Column 'basis functions'. Some of the most relevant basis functions which transform as the order parameter are listed. The symbols x,y,z indicate polar vectors in these directions, x, y, z indicate equivalent axial vectors. The tensor functions are [xx], [yy], [zz], [yz], [xz], [xy] and their combinations. If no such low rank tensors transform as basis function, higher rank tensors are given (e.g. x[yz] etc.] {} indicates the anti-symmetric product. As the order parameter(s) transform as the basis function(s), any combination of them can be a relevant part of the Landau potential (e.g. in $\bar{4}\,2m \rightarrow mm2$) the active rep. is B_2 with the two basis functions z and [xy] or any function with the same symmetry. Typical examples are the polarisation P_z and the strain e_6. The Landau potential is then:

$$G = \frac{1}{2} A(T-T_c)\, Q^2 + \frac{1}{4} B\, Q^4 + ... + \lambda_1\, P_z\, Q + \lambda_2\, e_6\, Q + ...$$

$$+ \frac{1}{2}\, \varepsilon\, P_z{}^2 + \frac{1}{2}\, c_{66}\, e_6^2$$

or, in thermodynamic equilibrium (Chapter 5 and 10):

$$G_Q = \frac{1}{2}\, \varepsilon^*\, P_z{}^2 + \frac{1}{2}\, c_{66}{}^*\, e_6^2 + \lambda^*\, P_z\, e_6 + ...$$

with possible anomalies in ε^*, $c_{66}{}^*$ and λ^*.

In cubic systems, it is convenient to use the symmetry adapted strains (see Chapter 8.2) for the E_g representation and the non-active A representation. These strains are (o stands for 'orthorhombic' and t stands for 'tetragonal'):

$$e_{vol} = e_1 + e_2 + e_3 \qquad\qquad \text{(A-representation)}$$

$$e_o = e_1 - e_2$$
$$\qquad\qquad\qquad\qquad \text{(E-representation)}$$
$$e_t = \frac{1}{\sqrt{3}}\, (2e_3 - e_1 - e_2)$$

The equivalent elastic constants are:

$$C_a = \frac{1}{3}(C_{11} + 2\, C_{12})$$

$$C_{o,t} = \frac{1}{2}(C_{11} - C_{12})$$

Table 1b   *The Form Of The Landau Potentials For Equitranslational Phase Transitions*

Landau Potential	Groups (irred. representations)
$L = 1/2\ A(T-T_c)Q^2 + 1/4\ BQ^4 + 1/6\ CQ^6$	all triclinic, monoclinic and orthorhombic systems
$L = 1/2\ A(T-T_c)(Q_1^2+Q_2^2)$ $+ 1/4\ B(Q_1^2+Q_2^2)^2 \cos\{4\ \text{arc tg}\ Q_1/Q_2\}$ $+ 1/4\ B'(Q_1^2+Q_2^2)^2 \sin\{4\ \text{arc tg}\ Q_1/Q_2\}$ $+ 1/4\ B''(Q_1^2+Q_2^2)^2$ $+ 1/6\ C\ (Q_1^2+Q_2^2)^3$ $+ 1/6\ C'(Q_1^2+Q_2^2)^3 \cos\{4\ \text{arc tg}\ Q_1/Q_2\}$ $+ 1/6\ C''(Q_1^2+Q_2^2)^3 \sin\{4\ \text{arc tg}\ Q_1/Q_2\}$	$4(E), \bar{4}(E), 4/m(E_g), 4/m(E_u)$
$L = 1/2\ A(T-T_c)(Q_1^2+Q_2^2)$ $+ 1/4\ B(Q_1^2+Q_2^2)^2 \cos\{4\ \text{arc tg}\ Q_1/Q_2\}$ $+ 1/4\ B'(Q_1^2+Q_2^2)^2$ $+ 1/6\ C'(Q_1^2+Q_2^2)^3$ $+ 1/6\ C''(Q_1^2+Q_2^2)^3 \cos\{4\ \text{arc tg}\ Q_1/Q_2\}$	$422(E), 4mm(E), \bar{4}2m(E), 4/mmm(E_g),$ $4/mmm(E_u)$
$L = 1/2\ A(T-T_c)(Q_1^2+Q_2^2)$ $+ 1/3\ B(Q_1^2+Q_2^2)^{3/2} \cos\{3\ \text{arc tg}\ Q_1/Q_2\}$ $+ 1/3\ B'(Q_1^2+Q_2^2)^{3/2}\sin\{3\ \text{arc tg}\ Q_1/Q_2\}$ $+ 1/4\ C\ (Q_1^2+Q_2^2)^2$ $+ 1/5\ D(Q_1^2+Q_2^2)^{5/2} \cos\{3\ \text{arc tg}\ Q_1/Q_2\}$ $+ 1/5\ D'(Q_1^2+Q_2^2)^{5/2}\sin\{3\ \text{arc tg}\ Q_1/Q_2\}$ $+ 1/6\ E\ (Q_1^2+Q_2^2)^3$ $+ 1/6\ E'(Q_1^2+Q_2^2)^3 \cos^2\{3\ \text{arc tg}\ Q_1/Q_2\}$ $+ 1/6\ E''(Q_1^2+Q_2^2)^3 \sin^2\{3\ \text{arc tg}\ Q_1/Q_2\}$ $+ 1/6\ E'''(Q_1^2+Q_2^2)^3 \sin\{6\ \text{arc tg}\ Q_1/Q_2\}$	$3(E), \bar{3}(E_g), 6(E_1), \bar{6}(E'), 6/m(E_{1g}), 23(E),$ $m3(E_g)$
$L = 1/2\ A(T-T_c)(Q_1^2+Q_2^2)$ $+ 1/3\ B(Q_1^2+Q_2^2)^{3/2} \cos\{3\ \text{arc tg}\ Q_1/Q_2\}$ $+ 1/4\ C\ (Q_1^2+Q_2^2)^2$ $+ 1/5\ D(Q_1^2+Q_2^2)^{5/2} \cos\{3\ \text{arc tg}\ Q_1/Q_2\}$ $+ 1/6\ E\ (Q_1^2+Q_2^2)^3$ $+ 1/6\ E'(Q_1^2+Q_2^2)^3 \cos^2\{3\ \text{arc tg}\ Q_1/Q_2\}$	$32(E), \bar{3}m(E), 3m(E_g), 622(E_2), 6mm(E_2),$ $\bar{6}m2(E'), 6/mmm(E_{2g}), 432(E), \bar{4}3m(E),$ $m3m(E_g)$
$L = 1/2\ A(T-T_c)(Q_1^2+Q_2^2)$ $+ 1/4\ B(Q_1^2+Q_2^2)^2$ $+ 1/6\ C\ (Q_1^2+Q_2^2)^3\cos\{6\ \text{arc tg}\ Q_1/Q_2\}$ $+ 1/6\ C'(Q_1^2+Q_2^2)^3\sin\{6\ \text{arc tg}\ Q_1/Q_2\}$ $+ 1/6\ C''\ (Q_1^2+Q_2^2)^3$	$3(E_u), 6(E_2), \bar{6}(E''), 6/m(E_{1u}), 6/m(E_{2g}),$ $6/m(E_{2u}), m3(E_u)$

Table 1b   *Cont'd*

Landau Potential	Groups (irred. representations)
$L = 1/2\ A(T-T_c)(Q_1^2+Q_2^2+Q_3^2)$ $+ 1/3\ B\ Q_1Q_2Q_3$ $+ 1/4\ C\ (Q_1^4+Q_2^4+Q_3^4)$ $+ 1/4\ C'(Q_1^2+Q_2^2+Q_3^2)^2$ $+ 1/6\ D\{Q_1^4(Q_2^2-Q_3^2)$ $\quad + Q_2^4(Q_3^2-Q_1^2)$ $\quad + Q_3^4(Q_1^2-Q_2^2)\}$ $+ 1/6\ D'(Q_1^2+Q_2^2+Q_3^2)^3$ $+ 1/6\ D''\ (Q_1Q_2Q_3)^2$ $+ 1/6\ D'''(Q_1^2+Q_2^2+Q_3^2)(Q_1^4+Q_2^4+Q_3^4)$	$23(T), m3(T_g)$
$L = 1/2\ A(T-T_c)(Q_1^2+Q_2^2+Q_3^2)$ $+ 1/4\ B\ (Q_1^4+Q_2^4+Q_3^4)$ $+ 1/4\ B'\ (Q_1^2+Q_2^2+Q_3^2)^2$ $+ 1/6\ C\ (Q_1Q_2Q_3)^2$ $+ 1/6\ C'\ \{Q_1^4(Q_2^2-Q_3^2)$ $\quad + Q_2^4(Q_3^2-Q_1^2)$ $\quad + Q_3^4(Q_1^2-Q_2^2)\}$ $+ 1/6\ C''\ (Q_1^2+Q_2^2+Q_3^2)^3$ $+ 1/6\ C'''(Q_1^2+Q_2^2+Q_3^2)(Q_1^4+Q_2^4+Q_3^4)$	$m3(T_u)$
$L = 1/2\ A(T-T_c)(Q_1^2+Q_2^2+Q_3^2)$ $+ 1/4\ B\ (Q_1^4+Q_2^4+Q_3^4)$ $+ 1/4\ B'\ (Q_1^2+Q_2^2+Q_3^2)^2$ $+ 1/6\ C\ (Q_1Q_2Q_3)^2$ $+ 1/6\ C'\ (Q_1^2+Q_2^2+Q_3^2)^3$ $+ 1/6\ C''\ (Q_1^2+Q_2^2+Q_3^2)(Q_1^4+Q_2^4+Q_3^4)$	$432(T_1), 43m(T_1), m3m(T_{1g}), m3m(T_{1u}),$ $m3m(T_{2u})$
$L = 1/2\ A(T-T_c)(Q_1^2+Q_2^2+Q_3^2)$ $+ 1/3\ B\ Q_1Q_2Q_3$ $+ 1/4\ C(Q_1^4+Q_2^4+Q_3^4)$ $+ 1/4\ C'\ (Q_1^2+Q_2^2+Q_3^2)^2$ $+ 1/6\ D\ Q_1Q_2Q_3\ (Q_1^2+Q_2^2+Q_3^2)$ $+ 1/6\ E\ (Q_1Q_2Q_3)^2$ $+ 1/6\ E'\ (Q_1^2+Q_2^2+Q_3^2)^3$ $+ 1/6\ E''\ (Q_1^2+Q_2^2+Q_3^2)(Q_1^4+Q_2^4+Q_3^4)$	$432(T_2), 43m(T_2), m3m(T_{2g})$

List of all Landau potentials for equitranslational phase transitions. The order parameter has two components for all E representations and three components for all T representations. The lowest order terms up to sixth order are given, no coupling between different order parameters is considered in this table. The group symbols refer to the high-symmetry form, the irreducible representations are the same as in Table 1a. The Landau potentials are listed using trigonometrical functions, where convenient, showing the inherent symmetry of the potentials when the components of the order parameter are interchanged.

TABLE 2

Table 2 Compilation of ferroelastic and co-elastic materials, and the known transition temperatures. In some cases the known magnitude of the spontaneous strain and the symmetries of the phases are given. The references are listed at the end of the table.

Compound	$T_c(K)$	Other Data	References
$Ag_3(AsS_3)$	29 22		Gorin Yu.F. et al. (1985)
AlF_3	725	$e_4 = 0.02$	Mogus-Milankovic A. et al. (1985), Ravez J. and Mogus-Milankovic A. (1985)
As_2O_5	578	$e_s = 0.023$ $P4_12_12 \rightarrow P2_12_12_1$	Redfern S.A.T. and Salje, E. (1988), Bismayer U. et al. (1986), Salje E. et al. (1987), Schmahl W. and Redfern S.A.T. (1988)
$BaMnF_4$	250	$A2_1am \rightarrow IC$	Saint-Gregoire P. et al. (1986)
$(Ba_2NaNb_5O_{15})$	853 573 543 105	$4/mmm \rightarrow$ $4mm \rightarrow$ $mm2$	Hebbache M. et al. (1984), Manolikas C. (1981), Schneck J. et al. (1982), Schneck J. and Denoyer F. (1981), Gumenyuk A.F. et al. (1987), Verwerft. M. et al. (1989)
$((CH_3)_3NCH_2COO \cdot H_3BO_3)$	142		Haussühl, S. (1984)
$BaTiO_3$	393		Flocken J.W. et al. (1985), Flerova S.A. and Taran V.G. (1979)
$(C_6H_5CO)_2$	83	$P3_121 \rightarrow C_2$ or $P3_221 \rightarrow C_2$	Vacher R. et al. (1981), Yoshihara A. et al. (1985), Moore D.R. et al. (1983)
Bismuth Halides	727 684 395		Barbier P. et al. (1984)
$BiNbO_4$	633		David W.I.F. (1983c)
$BiTaO_4$	843		
$BiVO_4$	520- 530	$I2/a \rightarrow I4_1/a$	Avakyants C.P. et al. (1982,1985), Baran N.P. et al. (1987), Boulesteix C. et al. (1986), David W.I.F. (1983a,b), David W.I.F. and Wood. I.G. (1983a,b), Dudnik E.F. et al. (1979), Gene V.V. and Shul'ga V.I. (1982), Lim A.-R. et al. (1988), Mnushkina I.E and Dudnik, E.F. (1982), Wainer L.S. et al. (1981), Hagen R.M. (1982)

Table 2   *Cont'd*

$C_6H_5NH_3Br$	300		Hattori A. et al. (1980), Sawada A. et al. (1980)
$(CH_3CH_2NH_3)_2CuCl_4$	362 349		Etxebarria T. et al. (1985)
$CaBr_2$	778	$P4_2/mnm \rightarrow Pnnm$	Bärnighausen H. et al. (1984)
$CaCl_2$	490	$P4_2/mnm \rightarrow Pnnm$	Unruh H.G. et al. (1991)
$CaTe_2O_5$	1083		Sadovskaya L.J. et al. (1983)
$(C_2H_5NH_3)_2CuCl_4$	361 349	$4/mmm \rightarrow mmm$	Tello M.J. et al. (1981)
$C_3N_3H_3$ S-triazine	200	$R\bar{3}c \rightarrow C2/c$	Luty T. and Van der Avoird A. (1983), Rae A.I.M. (1982), Raich J.C. et al. (1982, 1983), Heilman I.U. et al. (1979), Smith J.H. and Rae A.I.M. (1978a,b), Raich J.C. and Bernstein E.R. (1982)
$C_{18}H_{14}$ (p-Terphenyl)	193	$2/m \rightarrow \bar{1}$	Ecolivet C. et al. (1984)
$Cd_2Nb_2O_7$	196 205		Kolpakova N.N. et al. (1987)
$CeNbO_4$	903 196		Kukueva L.L. et al. (1984)
CeP_5O_{14}	128		Ting C. and Guang-yan H. (1986)
$(n\text{-}C_3H_5NH_3)_2ZnCl_4$	310	$Pnma \rightarrow P2_1/n$	Zuniga F.J. et al. (1982)
$[Co(H_2O)_6]GeF_6$	338	$\bar{3}m \rightarrow 2/m$	Hrabanski R. (1988)
CrF_3	1240	$e_s = 0.054$	Ravez J. et al. (1984), Mogus-Milankovic A et al. (1985)
CsH_2PO_4	503 153		Yoshida H et al. (1984a,b)
$CsSc(MoO_4)_2$	16		Fomin V.I. and Nesterenko N.M. (1986), Otko A.I. et al. (1984), Gurskas A.A. et al. (1985,1987)
Cs_2CrO_4	953		Dudnik E.F. (1977)
$CsDSO_4$	412		Dzhavadov N. & Plakida N.M. (1986)
$CsDy(MoO_4)_2$	42 59		Anders E.E. et al. (1985), Hamarda S. et al. (1989), Zuyagin A. (1983), Gurskas A.A. et al. (1983,1985), Skorobogatova I.V. et al. (1983,1986)

Table 2   *Cont'd*

Cs_2RbDyF_6	251	$Fm3m \rightarrow I4/m$	Aleksandrov K.S. et al. (1985,1987), Flerov I.N. et al. (1985), Skorobogatova I.V. et al. (1983,1986)
	205	$\rightarrow B2/m$	
		$\rightarrow P\bar{1}$	
	195		
Cs_2KDyF_6	169		Flerov I.N. et al. (1983)
Cs_2NaDyF_6	360		Flerov I.N. et al. (1983)
$Cs_2NaPrCl_6$	152		Flerov I.N. et al. (1983), Aleksandrov K.S. et al. (1984b)
$Cs_2NaNdCl_6$	132		Flerov I.N. et al. (1983), Aleksandrov K.S. et al. (1984b)
$Cs_2NaBiCl_6$	100		Flerov I.N. et al. (1983), Aleksandrov K.S. et al. (1984b)
$CsHSO_4$	414		Dzhavadov N. and Plakida N.M. (1986), Plakida N.M. and Shakhmatov V.S. (1987)
$CsHSeO_4$	398		Dzhavadov N. and Plakida N.M. (1986)
$CsLiSO_4$	202		Aleksandrov K.S. et al. (1980), Anistratov A. et al. (1982,1983), Morais P.C. et al. (1984)
$CsIO_4$	323		Al-Dhakhir T.A. et al. (1988)
Cs_2HgBr_4	243 230 165		Plesko S. et al. (1981), Vlokh O.G et al. (1988)
Cs_3BiCl_6	395 684 727		Von der Muhll R. et al. (1982)
$CsHSeO_4$	401		Yokota, S. and Makita Y. (1982), Yokota S. (1982)
$Cs_{2-x}K_{1+x}BiCl_6$	413- 583*		Barbier P. et al. (1982)
$CsPbCl_3$	320- 310		Chabin M. and Gilletta F. (1980a and b)
$CsLu(MoO_4)_2$	90		Gurskas A.A. et al. (1983)
$CsLu(WO_4)_2$	190		Gurskas A.A. et al. (1983
$CsHo(MoO_4)_2$	11		Skorobogatova I.V. et al. (1983)
DCN	160	$I4mm \rightarrow Imm2$	Mackenzie G.A. and Pawley G.S. (1979)

Table 2   *Cont'd*

DySb	9.5		Moran T.J. et al. (1973)
DyVO$_4$	15		Kaplan M.D. (1988), Sivardiere J. (1972) Melcher R.L. and Scott B.A. (1972), Sandercock J.R. et al. (1972)
DyNbO4	1143		Kukueva L.L. et al. (1984)
EuP$_5$O$_{14}$	159		Ting C. and Guang-yan H. (1986)
EuNbO$_4$	1103		Kukueva L.L. et al. (1984)
ErNbO$_4$	1083		Kukueva L.L. et al. (1984)
FeF$_3$	640	$e_s = 0.031$	Ravez J. and Mogus-Milankovic A. (1985), Mogus-Milankovic A. et al. (1985)
GaF$_3$	175		Ravez J. and Mogus-Milankovic A. (1985), Mogus-Milankovic A. et al. (1985)
Germanium hydrosodalite	166		Detinich V.A. et al. (1986)
GdNbO4	1093		Kukueva L.L. et al. (1984)
GdP$_5$O$_{14}$	175		Ting C. and Guang-yan H. (1986)
Gd$_2$(MoO$_4$)$_3$	432		Sakhnenko V.P. and Timonin P.N. (1983), Dvorak V. (1971), Ya-Gu W. et al. (1985), Gridnev S.A. et al. (1983) Bolshakova N.N. et al. (1983), Flerova S.A. and Taran V.G. (1979), Suzuki I. and Ishibashi Y. (1987)
H$_2$(UO$_2$)$_2$(AsO$_4$)$_2\cdot$8H$_2$O	293	4/mmm→2/m	Lanze de Dussel H. et al. (1981)
H$_2$C$_4$O$_4$	371		Matsushita E. and Matsubara T. (1981)
HCN	170	I4mm→Imm2 $e_s\approx0.10$	Dulmage W.J. and Lipscomb W.N. (1951), Mackenzie G.A. and Pawley G.S. (1979), Rae A.I.M. (1972)
Hg$_2$(Cl$_x$Br$_{1-x}$)$_2$	130-180*		Barta C. et al. (1979,1982), Benoit J.P. et al. (1982), Dobrzhanskii G.F. et al. (1983), Kaplyanskii, A.A. et al. (1979)
Hg$_2$Br$_2$	143	I4/mmm→Bbmm	Dobrzhanskii G.F.et al. (1983)

Table 2   *Cont'd*

Hg_2Cl_2	185	I4/mmm→Bbmm	Dobrzhanskii G.F.et al. (1983), Cao Xuan An et al. (1977)
Hg_2I_2	-18		Barta C.et.al.(1985)
$HoNbO_4$	1093		Kukueva L.L. et al. (1984)
9-hydroxyphenalenone	385 380 255		Svensson C. and Abrahams S.C. (1984)
InF_3	655	$e_s = 0.0558$	Ravez J. and Mogus-Milankovic A. (1985)
In-23 at % Th	286	$e_s \geq 0.017$	Saunders G.A. et al. (1986)
In-Tl	240-320		Parfenov O.E. and Chernyshov A.A. (1988), Pouget J. (1989)
In-Tl alloys 25% Tl In-Tl alloys 27% Tl	196 127		Gunton D.J and Saunders G.A. (1974)
$K_2Cd_2(SO_4)_3$	430	$P2_13→P2_12_12_1$	Hikita T. et al. (1977)
$(KBr)_{1-x}:(KCN)_x$ x = 0.73	112		Knorr K. et al. (1985,1986)
KCN	168	Fm3m→Immm $e_1 = -0.08$ $e_2 = 0.14$ $e_3 = -0.05$	Ivliev M.P. and Sakhnenko V.P. (1979), Hassühl S. (1973), Ohno E. (1987)
$KClO_3$	523	Pcmn→P2_1/m $e_s = 0.257$	Wadhawan V.K. (1980)
$KCaF_3$	~560 ~551		Bulou A. et al. (1980)
KD_2PO_4	220		Reese R.L. et al. (1973)
$KD_3(SeO_3)_2$	296	Pbcn→P2_1/b	Makita Y. et al. (1977), Schneider V.E. et al. (1987)
K_2CrO_4	893		Dudnik E.F. (1977)
K-feldspar	723	C2m→C$\bar{1}$	Harris M.J. et al. (1989)
KIO_2F_2	298		Abrahams S.C. and Bernstein J.L. (1976)
K_2NaDyF_6	190		Flerov I.N. et al. (1985)
$K_2Hg(CN)_4$	110		Wong P.T.T. (1980)
$K_3H(SeO_4)_2$	390	$e_5 = -5.1 \times 10^{-3}$ R$\bar{3}$m→A2/a	Yokota S. et al. (1982)

Table 2   *Cont'd*

$KMnF_3$	186		Shapiro S.M. et al. (1972)
K_3NbOF_6	283		Fouad M. et al. (1987)
K_3TaOF_6	310		Fouad M. et al. (1987)
K_2TeBr_6	434 400		Abrahams S.C. et al. (1984)
$K_4Zn(MoO_4)_3$	643		Klevtsova R.F. et al. (1986)
$KFe(MoO_4)_2$	312 139		Krainyuk G.G. et al. (1983a,b), Otko A.I. et al. (1979,1983,1984), Smolenskii G.A. et al.
$KIn(WO_4)_2$	454		(1979a,1980)
KH_2PO_4	122	$\bar{4}2m \rightarrow mm2$	Bastie P. and Becker P. (1984), Brody E.M. and Cummins H.Z. (1974)
$KH_3(SeO_3)_2$	211	$Pbcn \rightarrow P2_1/b$	Blinn R. et al. (1980), Darinskii B.M. et al .(1983), Gridnev S.A. and Shuvalov L.A. (1983), Gridnev S.A. et al. (1979a,b,1980), Ivanov N.R. (1979), Sol'tsas R.Kh. and Schneider V.E. (1983), Sorge, G. et al. (1981,1982a,b), Tanaka M.and Tatswzalis I (1983), Yamada Y. et al. (1983), Zapart W. and Zapart M.B. (1988), Makita Y. et al. (1977)
$KLiSO_4$	190	$P6_3mc \rightarrow Cmc2_1$	Zeks B. et al. (1984)
$KSCN$	415	$I4/mcm \rightarrow Pbcm$	Kroupa J. et al. (1988)
$KSc(MoO_4)_2$	258 181	$P\bar{3}m1 \rightarrow C2/m$ $\rightarrow C2/c$	Fomin V.I. and Nesterenko N.M. (1986), Gurskas A.A. et al. (1985,1987), Otko A.I. et al. (1984), Pelikh L.N. (1983a,b), Zapart M.B. and Zapart W. (1988)Zapart M.B. et al.(1982)
$KTaO_3{:}Li$	126 116		Ivliev M.P. and Sakhnenko V.P. (1986)
$K_2Zn_2(SO_4)_3$	138 75		Hikita T.et al.(1977)
$K_2Mn_2(SO_4)$	193	$P2_13 \rightarrow P2_12_12_1$	Hikita T et al.(1977)

Table 2   *Cont'd*

LaP_5O_{14}	393	Pmna→$P2_1$/b	Errandonea G. and Savary H. (1981), Errandonea G. (1980), Ting C. and Guang-yan H. (1987), Fox D.L. et al. (1976), Toledano, J.C. et al. (1976)Toledano P et al(1983)
$La_{1-x}Nd_xP_5O_{14}$(LNPP)	413	mmm→2/m	Suumi K.A. et al. (1980), Yening W. et al. (1987)Wenyuan S.et al.(1985)
$LaNbO_4$	792	$I4_1$/a→I2/b	Kukueva L.L. et al. (1984)Ishibashi Y. et al.(1988)Dudnik E.F. and Kiosse G.A.(1983)Toledano P.et al.(1983)
$LiNbO_3$	1470	R3c→R$\bar{3}$c	David W.I.F. (1983e)
$LiNH_4SO_4$	460 285	mmm→mm2 →2/m	Mroz B. et al. (1989)
$LiNH_4C_4H_4O_6$-H_2O(LAT)	98		Abe R. and Matsuda T. (1981), Sawada A. and Nakamura T. (1982)Sawada A. et al. (1977), Udagawa K. et al. (1978)
$LiRb_5(SO_4)_3$·$1.5H_2SO_4$	132	4mm→mm2	Mroz B. et al. (1988)Wolejuo T.et al.(1988)
$LiKSO_4$	940 700		Kassem M.E.M. and Mroz B. (1983), Krajewski T. et al. (1985)Balagurov A.M. et al.(1986)Xuxin Z et al.(1986)
$LiRb_4H(SO_4)_3$·H_2SO_4	137	4mm→mm2	Hempel H. et al. (1988)
$LiTaO_3$	895		David W.I.F. (1983e).
$LiCsSO_4$	202	Pcmn→$P2_1$/n	Asaki T. and Haebe K. (1988), Huang S.-J. and Yu J.-T.(1987)Ozeki H. and Tuszynski J.A. et al.(1988) , Pakulski G. et al. (1983)
$LnNbO_4$	1088		Kukueva L.L. et al. (1984)
$[Mg(H_2O)_6]SiFe_6$	298	R$\bar{3}$m→$P2_1$/c	Syoyama S. and Osaki K. (1972)
γ-Mn	420-470		Parfenov O.E. and Chernyshov A.A. (1988)Harley R.T.et al.(1978)
$(N(CH_3)_4)_2CuCl_4$	299 293 265		Gomez-Cuevas A. et al. (1981,1983), Martin B. et al. (1982), Sawada A. et al. (1980b,1981), Sugiyama J. et al. (1980)

Table 2   *Cont'd*

NaOH	513		Bleif H. et al. (1971)
$Na_3MoO_3F_3$	530 130	$Fm3m \rightarrow P2_1$	Chaminade J.P. et al. (1986)
$NaH_3(SeO_3)_2$	194	$2/m \rightarrow \bar{1}$ $e_s = 0.0029$	Gridnev S.A. et al. (1985)
NaN_3	293	$R3m \rightarrow C2/m$ $e_s \approx 0.05,$ $e_1 - e_2 = 0.02$	Hirotsu S. (1983), Stevens, E.D. and Hope H. (1977), Pringle G.E. and Noakes D.E. (1968), Aghadee S.R. and Rae A.I.M. (1984)
Na_2CO_3	863 633 130	$P6_3/mmc \rightarrow C2/m$	Midorikawa M. et al. (1980)
$Na_5Ti_3O_3F_{11}$	765	$e_s = 0.075$	Yacoubi A. et al. (1987)
$NaNDSeO_42D_2O$	180		Sandler Yu.M. et al.(1978)
$Na_5W_3O_9F_5$	800 530	$e_s = 0.067$	Ravez J. et al. (1979,1983)
Na-feldspar	1250	$C2/m \rightarrow C\bar{1}$ $e_s = 0.045$	Salje E. et al.(1985)
$Na_5Al_3F_{14}$	150	$P4/mnc \rightarrow P2_1/n$	Rocquet P. and Couzi M. (1985)
NaCN	284		Haussühl S. et al. (1977)Ohno E.(1987)Rowe J.M. et al.(1975)
Nb_3Sn	49		Rehwald W. (1968)
Nd_2O_3	293	$\bar{3}n \rightarrow 2/m$ $e_s = 0.04$	Ben Salem M. et al. (1984)Boulesteix C. et al.(1986)
NdP_5O_{14}	419	$Pncm \rightarrow P2_1/c$	Errandonea G. (1980), Fox D.L. et al. (1976)Dudnik E.F. and Kiosse G.A.(1983)
$NdNbO_4$	1003	$I4_1/a \rightarrow I2/a$	Kukueva L.L. et al. (1984)Ishibashi Y. et at.(1988)
$NdTaO_4$	1601	$\varepsilon_s = 0.008$ $\bar{4}2m \rightarrow 2/a$	David, W.I.F. (1983)
$Nd_2(MoO_4)_3$	1233		Prewitt C.T. (1970)
NH_4BeF_3	526 333 251		Yoshida H. et al. (1984a,b)
NH_4HBeF_4	333		Yoshida H. et al. (1984b)

Table 2   *Cont'd*

NH_4HSeO_4	417		Dzhavadov N.andPlakida N.M.(1986)
$(NH_4)_3H(SeO_4)_2$	302	$R\bar{3}m \rightarrow A2/a$ $e_s = 0.03$	Kishimoto T. et al. (1987), Alexandrov K.S. et al.(1983)
$(NH_4)_3H(SO_4)$	413 265 141 133		Osaka T. et al. (1977)
$(NH_4)2Cd_2(SO_4)_3$	95	$P2_13 \rightarrow P2_1$	Hikita T. et al(1977)
$NiCr_2O_4$	300		Kino Y. et al. (1973)
NiF_2	$P_c=$ 18.3 Kbar		Jorgensen J.D. et al. (1978)
phenothiazine	251	$Pbnm \rightarrow P2_1/n$	Nakayama H. et al. (1985)
$Pb_3(PO_4)_2$	453	$R\bar{3}m \rightarrow C2/c$	Bismayer U. et al. (1982), Benoit J.P. et al. (1981), Darlington C.N.W. (1983), Roucau C. et al. (1983), Salje E. and Wruck B. (1983), Torres J. et al. (1980a,b,1981,1982), Vagin S.V.(1979), Smirnov P.S. et al. (1979), Wood I.G.et al.(1980), Dudnik E.F. and Nepochatenko V.A.(1980a,b), Salje E. and Devarajan V. (1981)
$PbK_2(MoO_4)_2$	453	$R\bar{3}m \rightarrow P2/m$	Dudnik E.F. and Kiosse G.A.(1983)
$Pb_3(VO_4)_2$	370 270	$R\bar{3}m \rightarrow P2_1/c$ $\rightarrow P2_1$ or Pc	Von Hodenberg R. and Salje E. (1977), Manolikas C. and Amelinckx S.(1980)
$Pb_3(AsO_4)_2$	548 333	$R\bar{3}m \rightarrow C2/c$ $\rightarrow P2_1/c$ $e_s = 0.02$	Bismayer U. and Salje E. (1981), Bismayer U. et al.(1982), Dudnik E.F. and Sushko S.A. (1983), Kiosse G.A. et al. (1981)
$Pb_3(P_{1-x}As_xO_4)_2$	250-508*	$R\bar{3}m \rightarrow C2/c \rightarrow P2_1/c$	Bismayer U. et al. (1986)Salje et al.(1983)
$PbO(\alpha)$	200	$P4/nmm \rightarrow Cmma$	Boher P. et al. (1985)
$Pb_8V_2O_{13}$	432		Dudnik E.F.and Kolesov I.S. (1980), Kiosse G.A. et al. (1982), Dudnik E.F.et al.(1980)
$PbBi_2Nb_2O_9$	598		Delaporte O. et al. (1982)
$PrNbO_4$	953		Kukueva L.L. et al. (1984)

Table 2   *Cont'd*

PrAlO$_3$	119		Kjems J.K. et al. (1973),
	151		Fleury P.A. et al.(1974)
PrCu$_2$	≤10		Andres K. et al. (1976)
PrP$_5$O$_{14}$	138		Ting C. and Guang-yan H. (1986)
RbSc(MoO$_4$)$_2$	258	P$\bar{3}$m/→C2/m	Zapart M.B.and Zapart W.(1988)
	181	→C2/C	
Rb$_2$KDyF$_6$	381	Fm3m→B2/m	Flerov I.N. et al. (1985),
			Alexsandrov K.S. et al.(1984b)
Rb$_2$NaDyF$_6$	190	Fm3m→I4/m	Flerov I.N. et al. (1985),
			Alexsandrov K.S. et al.(1984b)
Rb$_2$NaHoF$_6$	172		Alexandrov K.S. et al.(1984b)
Rb$_2$Cd$_2$(SO$_4$)$_3$	129	P2$_1$3→P2$_1$	Hikita T.et al.(1977)
	103	→P$_1$	
Rb$_3$H(SeO$_4$)$_2$	447	R$\bar{3}$m/→C2/c	Plakida N.M. and Salejda W.
			(1988a,b), Dzhavadov N. and
			Plakida N.M.(1986)
RbCaF$_3$	194	Pm3m→I4/mcm	Ridou C. et al. (1980), Jex H. et al.
			(1980), Rousseau et al. (1980),
			Flocken J.W. et al (1985)
RbCN	132		Rowe J.M. et al. (1984), Ehrhardt
			K.D. et al. (1983), Ohno E.(1987)
Rb$_2$CdCl$_4$	143		Gorev M.V. et al. (1987)
RbCrO$_4$	923		Dudnik E.F. (1977)
RbMnCl$_3$	271	6mmm→2/m	Aleksandrov K.S. et al. (1984a),
			Gomez-Cuevas A. et al. (1984),
			Shchedrina N.V. (1980), Sandler
			Yu.M. and Serdobolskaja O.Yu.
			(1983)
RbD$_2$PO$_4$	317		Yoshida H. et al. (1984b)
	377		
RbH$_2$PO$_4$	144	$\bar{4}$2m→mm2	Bastie P. and Becker P. (1984)
RbAlF$_3$	559	P4/mmm→P4/mbm	Bulou A. et al. (1981), Bulou and
	282	→Pmmn	Nouet J. (1982) Kleeman W. et al.
			(1982)
RbAlF$_4$	553	P4/mmm→P4/mbm	Bulou and Nouet J. (1982)
	282	→Pmmn	
SF$_6$	96	Im3m→C2/m	Dolling G. et al. (1979) Cockcroft
		e$_3$=0.036	J.K. and Fitch A.N. (1988),
		e$_5$=0.04	Dove M.T. et al. (1988)

Table 2   *Cont'd*

Sb_5O_7I	481	$P6_3/m \rightarrow P2_1/C$	Grekov A.A. et al. (1981), Rehwald W. et al. (1980)
SiO_2 quartz	846	$P6_222 \rightarrow 1NC$ $\rightarrow P3_221$	Kihara K. (1990), Dolino G. et al. (1983), Salje E.K.H. et al. (1991)
$SmNbO_4$	1098		Kukueva L.L. et al. (1984)
SmP_5O_{14}	166		Ting C. and Guang -Yan H (1986)
$SrTe_2O_5$	908		Sadovskaya L.Ja. et al. (1983)
Sr_2SiO_4	367 357	$mmm \rightarrow 2/m$	Catti M. et al. (1983), Catti M. and Gazzoni G. (1983).
$SrTe_2O_5$			Salovskaya C.Ya. et al. (1981)
$Sr_2KNb_5O_{15}$	429		Manolikas C. (1981)
$SrTiO_3$	100		Shapiro S.M. et al. (1972)
TbP_5O_{14}	179		Errandonea G. (1980)
TeO_2	$P_{c=}$ 9Kbar		Peercy P.S. and Fritz I.J. (1974)Skelton E.F. et al.(1976)
$TbVO_4$	34		Sandercock J.R. et al. (1972)
TiF_3	340	$e_s = 0.014$	Ravez, J.and Mogus-Milankovic A. (1985), Mogus-Milankovic A. et al. (1985)
tanane	287	$\bar{4}2m \rightarrow mm2$	Bordeaux D. et al. (1973), Luty T. and van der Avoird A. (1983), Lajzerowicz J. et al. (1980)
$TlAlF_4$	514 435	$P4/mmm \rightarrow I4/mcm$ $\rightarrow I2/a$	Bulou A. and Nouet J. (1987)
$TlNO_3$	416 349	$Pm3m \rightarrow P3_1$ $\rightarrow Pnma$	Wadwahan V.K.and Somajazalu M.S. (1986)
TlH_2PO_4	357	$e_s = 0.023$	Yoshida H. et al. (1984b)
TiH_2	300		Kulikov N.I. and Tuguskev V.V. (1981)
$Tl_2Cd_2(SO_4)_3$	128 120 92	$P2_13 \rightarrow P2_1$ $\rightarrow P1$ $\rightarrow P2_12_12_1$	Hikita T.et al.(1977)
$TmCd$	32		Lüthi B.et al.(1973)
$TmVO_4$	2		Melcher R.L. et al. (1973)
$TSCC$	125		Smolenskii G.A. et al. (1979b)

Table 2   *Cont'd*

UO_2	31		Brandt O.G. and Walker C.T. (1967)
VF_3	770	$e_s = 0.038$	Ravez J. and Mogus-Milankovic A. (1985)
V_3Si	21		Parfenov O.E. and Chernysov A.A. (1988), Batterman B.W. and Barrett C.S. (1966)Toledano P. et al.
VO_2	340		Paquet D. and Leroux-Hagon P. (1979)
$YBa_2Cu_3O_7$	970	$e_s = 0.02$	Schmahl W.W.et al. (1989), Wadwahan V.K. and Glazer A.M. (1989), Salje E.K.H. and Parlinski K. (1991)
$YbNbO_4$	1098		Kukueva L.L. et al. (1984)
$YNbO_4$	1093		Kukueva L.L. et al. (1984)

Notes: IC = incommensurate phase; * = depending on composition

References for table 2

ABE, R. & MATSUDA, T. (1981). Microscopic origin of the ferroelastic phase transition in $LiNH_4C_4H_4O_6.H_2O$. *J. Phys.Soc.Japan.* **50(1)**, 163.

ABRAHAMS, S.C. & BERNSTEIN, J.L. (1976). Ferroelastic KIO_2F_2: crystal structure and ferroelastic transformation. *J. Chem Phys.* **64**, 3254.

ABRAHAMS, S.C., IHRINGER, J., MARSH, P. & NASSAU, K. (1984). Phase transition at 434K, independent strain coupling in second transitions at 400K, and thermal expansivity in ferroelastic K_2TeBr_6. *Chem. Phys.* **81**, 2082.

AGHADEE, S.R. & RAE, A.I.M., (1984). The low temperature phase of sodium azide. *Acta Cryst.* **B40**, 214.

AL-DHAKHIR, T.A., RAGHUNATHA-CHARY, B., BHAT, H.L. & NARAYANAN, P.S. (1988). Raman spectroscopic studies of ferroelastic cesium periodate. *Indian J. Pure Appl. Phys.* **26(2-3)**, 239.

ALEKSANDROV, K.S., BOVINA, A.F., VORONOV, V.N., GOREV, M.V., ISKORNEV, I.M., MELNIKOVA, S.V., MISJUL, S.V., PROKERT, F. & FLEROV, I.N. (1985). Ferroelastic phase transitions in elpasolites $A_2Br_3+X_6$. *Japan. J. Appl. Phys. Suppl.* **24** , 699.

ALEKSANDROV, K.S., KRUGLIK, A.I. & ZINENKO, V.I. (1983). Structural investigations of new families of crystals. *Bull. Acad. Sci. USSR, Phys. Ser.* **47(3)**, 103.

ALEKSANDROV, K.S., ZANIKOV, A.V. & KRUGLIK, A.I. (1984a). Thermodynamic characteristics of $RbMnCl_3$ ferroelastic. *Bull. Acad. Sci. USSR, Phys. Ser.* **48**, 131.

ALEKSANDROV, K.S., FLEROV, I.V., BOVINA, A.F., VORONOV, V.N., GOREV, M.V., MELNIKOVA, S.V., MISJUL, S.V (1984b). The study of phase transitions in single crystals with elpasolite structure. *Ferroelectrics* **54(1-4)**, 577.

ALEKSANDROV, K.S., ZHEREBTSOVA, L.I., ISKORNEV, I.M., KRUGLIK, A.I., ROZANOV, O.V. & FLEROV, I.N. (1980). Investigation of structural and physical properties of cesium-lithium double sulfate. *Sov. Phys. - Solid State* **22(12)**, 2150.

ALEKSANDROV, K.S., MELNIKOVA, S.V. & MISJUL, S.V. (1987). Successive phase transitions in ferroelastic Cs_2RbDyF_6 with elpasolite-type structure. *Phys. Status Solidi* **A104(2)**, 545.

ANDERS, E.E., VOLCHON, I.V., ZUYAGIN, A.I. & STARTSEV, S.V. (1985). Features of kinetic and thermodynamic characteristics of layered $CsDy(MoO_4)_2$ crystals in the region of ferroelastic transition. *Ferroelectrics* **64**, 335.

ANDRES, K.,WANG, P.S., WONG, Y.H., LUTHI, B.& OTT, H.R. (1976). In *Magnetism and Magnetic materials*- Ed.J.J.Becker, A.I.P. Conf. proc. No 34(AIP, New York, 1976), 222.

ANISTRATOV, A., ZANIKOV, A.M., KOT, L.A., STOLOVITSKAYA, I.N. & SHABANOVA, L.A. (1983). The properties of $CsLiO_4$ single crystals near Curie point. *Ferroelectrics* **48(1-3)**, 103.

ANISTRATOV, A., ZUMKOV, A.V., KOT, L.A., STOLOVITSKAYA, I.N. & SHABANOVA, L.A. (1982). Static properties of $CsLiSO_4$ near the ferroelastic phase transition. *Sov. Phys. - Solid State* **24(9)**, 1565.

ASAKI, T. & HAEBE, K. (1988). X-ray study of $LiCsSO_4$ in connection with its ferroelastic phase transition. *J. Phys. Soc. Japan.* **57(12)**, 4184.

AVAKYANTS, C.P., KISELEV, D.F. & CHERVYAKOV, A.V. (1985). Ferroelastic phase transition in $BiVO_4$. *Sov. Phys. - Crystallogr.* **30**, 595.

AVAKYANTS, C.P., ANTONENKO, A.M., DUDNIK, E.F., KISELEV, D.F., MRUSHKINA, I.E. & FIRSOVA, M.M. (1982). Elastic constants of ferroelastic $BiVO_4$. *Sov. Phys. - Solid State* **24(8)**, 1411.

BALAGUROV, A.M., POPE, N.C. & SAVENKO, B.N. (1986). Symmetry analysis of the low-temperature domain structure in ferroelastic $LiKSO_4$. *Phys. Status Solidi B* **134**, 457.

BARAN, N.P., BARCHUK, V.I. & GRACHEV, V.G. (1987). Peculiarities of EPR spectra under erosion of phase transition in imperfect crystals of intrinsic ferroelastic $BiVO_4$. *Ukr. Fiz. Zh.* **32**, 243. (In Russian).

BARBIER, P., DRACHE, M., MAIRESSE, G. & RAVEZ, J. (1982). Phase transitions in a $Cs_{2-x}K_{1+x}BiCl_6$ solid solution. *J. Solid State Chem.* **42(2)**, 130.

BARBIER, P., DRACHE, M., MAIRESSE, G. & RAVEZ, J. (1984). Properties of bismuth halides with $(NH_4)_4FeF_6$ or K_2NaAlF_6 related structure. *Ferroelectrics* **55(1-4)**, 113.

BÄRNIGHAUSEN, H., BOSSERT, W., AUSELMENT, B., (1984). A second-order phase transition of calcium bromide and its geometrical interpretation. *Acta Cryst.* **A 40**, C-96.

BARTA, C., KAPLYANSKII, A.A., MARKOV, YU.F. & MIROVITSKII, V.YU. (1985). Pressure-induced phase transition in virtual ferroelastic Hg_3I_2. *Sov. Phys. - Solid State* **27**, 1497.

BARTA, C., ZADOKHIN, B.S., KAPLYANSKII, A.A., MALKIN, B.Z., MARKOV, YU.F., MOROZOVA, O.V. & SAVCHENKO, B.A. (1979). Thermal expansion and thermodynamic potential of the non-intrinsic ferroelastic Hg_2Br_2. *Sov. Phys. Crystallogr.* **24(5)**, 608.

BARTA, C., ZADOKHIN, B.S., MARKOV, YU.F. & MOROZOVA, O.V. (1982). Spontaneous deformation and phase transition in mixed $Hg_2Cl_{1.2}Br_{0.8}$ crystals. *Sov. Phys.* **24(5)**, 867.

BASTIE, P. & BECKER, P. (1984). Gamma-ray diffraction in the vicinity of a ferroelastic transition: application to a 'real crystal' of RbH_2PO_4 and KH_2PO_4. *J. Phys. C* **17(2)**, 193.

BATTERMAN, B.W. & BARRETT, C.S. (1966). Low Temperature structural transformation in V_3Si. *Phys.Rev. B* **145** 296.

BEN SALEM, M., DORBEZ, R., YANGUI, B. & BOULESTEIX, C. (1984). Ferroelastic character and study by HREM of the mechanism of the hexagonal-monoclinic phase transition of rare earth sesquioxides. *Phil. Mag. A* **50**, 621.

BENOIT, J.P., HENNION, B. & LAMBERT, M. (1981). Dynamics of the ferroelastic phase transition of lead phosphate $(PO_4)_2Pb_3$. *Phase Transitions* **2(2)**, 103.

BENOIT, J.P., HAURET, G. & LEFEBVRE, J. (1982). Ferroelastic phase transition of Hg_2Cl_2. Study by neutron scattering; soft mode and central peak. *J. Phys.* **43(4)**, 641 (in French).

BISMAYER, U. & SALJE, E. (1981). Ferroelastic phases in $Pb_3(PO_4)_2$-$Pb_3(AsO_4)_2$ X-ray and optical experiments. *Acta Cryst. A* **37**, 145.

BISMAYER, U., SALJE, E., GLAZER, A.M. & COSIER, J. (1986). Effect of strain-induced order parameter coupling on the ferroelastic behaviour of lead phosphate arsenate. *Phase Trans.* **6**, 129.

BISMAYER, U., SALJE, E. & JOFFRIN, C. (1982). Re-investigation of the stepwise character of the ferroelastic phase transition in lead phosphate-arsenate, $Pb_3(PO_4)_2$-$Pb_3(AsO_4)_2$. *J. Phys.* **43(9)**, 1379.

BLEIF, H., DACHS, H., & KNORR, K. (1971). Diffuse X-ray scattering and change of lattice parameters of NaOH at transition from orthorhombic to monoclinic phase. *Sol. State Comm.* **9**, 1893.

BLINN, R., ZEKS, B., CHARLES, A.S. (1980). Microscopic theory of the ferroelastic transition in $KH_3(SeO_3)_2$. *Phys. Rev. B* **22(7)**, 3486.

BOHER, P., GARNIER, P., GAVARRI, J.R. & HEWAT, A.W. (1985). Alpha PbO quadratic monoxide. I. Description of the structural ferroelastic transition. *J. Solid State Chem.* **57**, 343 (In French).

BOLSHAKOVA, N.N., SOROKINA, I.I. & RUDYAK, V.M. (1983). Heating rate effect on domain structure realignment in gadolinium molybdate crystals. *Ferroelectrics* **48(1-3)**, 183.

BORDEAUX, D., BORNAREL, J., CAPIOMONT, A., LAJZEROWICZ-BONNETEAU, J., LAJZEROWICZ, J. & LEGRAND, J.F. (1973). New ferroelastic-ferroelectric compound - tanane. *Phys.Rev. Lett.* **31** 314.

BOULESTEIX, C., YANGUI, B., BEN SALEM, M., MANOLIKAS, C. & MELINCLIX, S.A. (1986). The orientation of interfaces between a prototype phase and its ferroelastic derivatives: theoretical and experimental studies. *J. de Phys.* **47**, 461.

BRANDT, O.G. & WALKER, C.T. (1967). Temperature dependence of elastic constants and thermal expansion for UO_2. *Phys.Rev.Lett.* **18** 11.

BRODY, E.M. & CUMMINS, H.Z. (1974). Brillouin scattering of elastic anomaly in ferroelectric KH_2PO_4. *Phys.Rev B* **9**, 179.

BULOU, A.& NOUET, J. (1982). Structural phase transition in ferroelastic $RbAlF4$.1-DSC, X-ray powder diffraction investigations and neutron powder profile refinement of the structure. *J.Phys. C.* **15** (2), 183.

BULOU, A. & NOUET, J. (1987). Structural phase transition in ferroelastic $TlAlF_4$: DSC investigations and structures determination by neutron powder profile refinement. *J. Phys. C* **20**, 2885.

BULOU, A., NOUET, J., HEWAT, A.W. & SCHÄFER, F.J. (1980). Structural phase transitions in $KCaF_3$-DSC-birefringence and neutron powder diffraction results. *Ferroelastics* **25** 375.

BULOU, A., NOUET, J., KLEEMAN, W., SCHAFER, W.J. & HEWAT, A.W. (1981). The structures and the improper ferroelastic phase transition of $RbAlF_4$. *Ferroelectrics* **36(1-4)**, 407.

CAO XUAN AN, HAURET G., & CHAPELLE, J.P. (1977). Brillouin scattering in Hg_2Cl_2. *Sol. State Comm* **24**, 443.

CATTI, M. & GAZZONI, G. (1983). The beta <==> alpha phase transition of $SrSiO_4$. II. X-ray and optical study, and ferroelasticity of the beta form. *Acta Crystallogr.* **39(6)**, 6679.

CATTI, M., GAZZONI, G., IVALDII, G. & ZANINI, G. (1983). The beta <==> alpha phase transition of $SrSiO_4$. I. order-disorder in the structure of the alpha form at 383K. *Acta Crystallogr.* **39(6)**, 674.

CHABIN, M. & GILLETTA, F. (1980a). Experimental investigation of the ferroelastic domain structure in cesium lead chloride in the monoclinic phase. *J. Appl. Crystallogr.* **13(6)**, 539.

CHABIN, M. & GILLETTA, F. (1980b). Theoretical investigation of the ferroelastic domain structure in cesium lead chloride in the monoclinic phase. *J. Appl. Crystallogr.* **13(6)**, 533.

CHAMINADE, J.P., CERVERA-MARCAL, M., RAVEZ, J. & HAGENMULLER, P. (1986). Ferroelastic and ferroelectric behaviour of the oxyfluoride $Na_3MoO_3F_3$. *Mater.Res.Bull.* **21**, 1209.

COCKCROFT, J.K. & FITCH, A.N., (1988). The solid phases of sulfur hexafluoride by powder neutron diffraction. *Z. Kristallogr.* **184**, 123

DARINSKI, B.M., NECHAED, V.N, PEREVOSNILIOV, A.M. (1983). The interaction between a dislocation and a domain in a ferroelastic. *Ferroelectrics* **48(1-3)**:17.

DARLINGTON, C.N.W. (1983). On the dynamics of low-temperature domains above T_c in lead phosphate. *Phase Transitions* **3(4)**, 283.

DAVID, W.I.F. (1983a). Ferroelastic phase transition in $BiVO_4$: III. Thermodynamics. *J. Phys. C* **16(26)**, 5093.

DAVID, W.I.F. (1983b). Ferroelastic phase transition in $BiVO_4$: IV. Relationships between spontaneous strain and acoustic properties. *J. Phys. C* **16(26)**, 5119.

DAVID, W.I.F. (1983c). Ferroelasticity in scheelite structures. *Solid State Chemistry Conf. Proc.*, 805.

DAVID, W.I.F. (1983d). Structural relationship between spontaneous strain and acoustic properties in ferroelastics. *J. Phys. C* **16(13)**, 2455.

DAVID, W.I.F. (1983e). Transition temperature - spontaneous strain - atomic displacement relationships in ferroelastics. *Mater.Res.Bull* **18(7)**, 809.

DAVID, W.I.F. & WOOD, I.G. (1983a). Ferroelastic phase transition in $BiVO_4$: V. Temperature dependence of Bi^{3+} displacement and spontaneous strain. *J. Phys. C* **16(26)**, 5127.

DAVID, W.I.F. & WOOD, I.G. (1983b). Ferroelastic phase transition in $BiVO_4$: VI. Some comments on the relationship between spontaneous deformation and domain walls in ferroelastics. *J. Phys. C* **16(26)**, 5149.

DELAPORTE, O., BARSCH, G.R., CROSS, L.E. & RYBA, E. (1982). Unusual ferroelastic behaviour in ferroelectric lead bismuth niobate ($PbBi_2Nb_2O_9$). *Ferroelectr. Lett. Sect.* **44(1)**, 1.

DETINICH, V.A., IVANOV, N.R. & GALITSKII, V.Yu. (1986). The ferroelastic phase transition and low-frequency dielectric relaxation in germanium hydrosodalite *Bull.Acad.Sci.USSR, Phys. Ser.* **50**, 173.

Appendix

DOBRZHANSKII, G.F., KAPLYANSKII, A.A., LIMONDE, M.V. & MARKOV, YU.F. (1983). A ferroelastic phase transition in $Hg_2(Cl_xBr_{1-x})_2$ crystals. *Ferroelectrics* **48(1-3)**, 69.

DOLINO, G., BACHHEIMER, J.P. & ZEYEN, C.M.E., (1983). Observation of an intermediate phase near the α-β transition of quartz by heat capacity and neutron scattering measurements. *Solid State Comm.* **45**, 295.

DOLLING, G., POWELL, B.M. & SEARS, V.F., (1979). Neutron-diffraction study of the plastic phases of polycrystalline SF_6 and CBr_4. Molec.Phys. **37**, 1859.

DOVE, M.T., POWELL, B.M., PAWLEY, G.S. & BARTELL, L., (1988). Monoclinic phase of SF_6 and the orientational ordering transition. *Molec. Phys.* **65**, 353.

DUDNIK, E.F. (1977). New ferroelastics K_2CrO_4, Rb_2CrO_4 and Cs_2CrO_4. *Soviet Physics Solid State* **19** 502.

DUDNIK, E.F. & KIOSSE, G.A. (1983). The structural peculiarities of some pure ferroelastics. *Ferroelectrics* **48**, 33.

DUDNIK, E.F., KIOSSE, G.A. & STOLPAKOVA, T.M. (1983). Features of the domain structure of ferroelastics of the type $NaFe(MoO_4)_2$. *Bull. Acad. Sci. USSR, Phys. Ser.* **47(3)**, 82.

DUDNIK, E.F. & KOLESOV, I.S. (1980). New ferroelastics $Pb_8V_2O_{13}$ and $Pb_8P_2O_{13}$. *Sov. Phys.* **22(4)**, 700.

DUDNIK, E.F., KOSELOV, I.S. & ORLOV, O.L. (1980). Domain structure and phase transitions in antiferroelectric ferroelastic lead orthovanadate. *Sov. Phys.* **22(6)**, 1084.

DUDNIK, E.F. & SUSHKO, S.A. (1983). Domain structure of the ferroelastic lead orthoarsenate. *Ferroelectrics* **48(1-3)**, 149.

DUDNIK, E.F. & NEPOCHATENKO, V.A. (1980a). Dielectric anomalies of ferroelastics - $Pb_3(PO_4)_2$. *Ukr.Fiz.Zh.* **25(4)**, 678. (In Russian).

DUDNIK, E.F. & NEPOCHATENKO, V.A. (1980b). Phase boundaries in the ferroelastic lead orthophosphate. *Sov.Phys.Crystallogr.* **25(5)**, 564.

DUDNIK, E.F., GENE, E.E. & MRUSHKINA, I.E. (1979). Ferroelastic phase transition in $BiVO_4$ single crystals. *Bull. Acad. Sci. USSR, Phys. Ser.* **43(8)**, 149.

DULMAGE, W.J. & LIPSCOMB, W.N., (1959). The crystal structure of hydrogen cyanide HCN. *Acta Cryst.* **4**, 330.

DVORAK, V(1971). A thermodynamic theory of galolinium molybdate. *Phys .Stat Sol.* **46**, 763 .

DZHAVADOV, N. & PLAKIDA, N.M. (1986). A ferroelastic phase transition in a proton supersonic conductor. *Bull. Acad. Sci. USSR, Phys. Ser.* **50(2)**, 97.

ECOLIVET, C., TOUDIS, B. & SANQUER, M. (1984). Acoustic anomalies at the improper ferroelastic phase transition of p-terphenyl. *Ferroelectrics* **54(1-4)**, 613.

ERHARDT, K.D., PRESS, W. & HEGER, G. (1983). Structure analysis of the disordered cubic phase of rubidium cyanide.*Acta Cryst.* B **39**, 171.

ERRANDONEA, G. (1980). Elastic and mechanical studies of the transition in LaP_5O_{14}: a continuous ferroelastic transition with a classical Landau-type behaviour. *Phys .Rev.B* **21(11)**, 5221.

ERRANDONEA, G. & SAVARY, H. (1981). Study of the ferroelastic transition of LaP_5O_{14} under high hydrostatic pressure. *Ferroelectrics* **36(1-4)**, 427.

ETXEBARRIA, T., FERNANDEZ, J., ARRIANDIAGA, M.A. & TELLO, M.J. (1985). Influence of the thermal expansion on the piezoelectric photoacoustic detection of ferro-paraelastic phase transition in $(CH_3CH_2NH_3)_2CuCl_4$. *J. Phys. C* **18**, L13.

FLEROV, I.N., BOVINA, A.F., VORONOV, V.N., GOREV, M.V., MISJUL, S.V., MELNIKOVA, S.V. & SHABANOVA, L.A. (1985). Ferroelastic phase transitions in elpasolites. *Ferroelectrics* **64**, 341.

FLEROV, I.N., GOREV, M.V. & ISKORNED, I.M. (1983). Calorimetric and dilatometric study of the ferroelastic phase transition in the elpasolites. *Ferroelectrics* **48(1-3)**, 97.

FLEROVA, S.A. & TARAN, V.G. (1979). Luminescence during mechanical switching of crystals having ferroelectric and ferroelastic properties. *Bull.Acad.Sci. USSR, Phys. Ser.* **43(8)**, 169.

FLEURY, P.A., LAZAY, P.D. & VAN UITER, T. (1974). Brillouin scattering evidence for a new phase transition in perovskite crystals, PRA103. *Phys. Rev. Lett.* **33**, 492.

FLOCKEN, J.W., GUENTHER, R.A., HARDY, J.R. & BOYER, L.L. (1985). First-principles study of structural instabilities in halide-based perovskites: competition between ferroelectricity and ferroelasticity. *Phys. Rev. B* **31**, 7252.

FOMIN, V.I. & NESTERENKO, N.M. (1986). First-order ferroelastic phase transition in $KSc(MoO_4)_2$. *Sov. Phys. - Crystallogr.* **31**, 485.

FOUAD, M., RAVEZ, J., CHAMINADE, J.P. & HAGENMULLER, P. (1987). Study of the phase transitions of new ferroelastic oxyfluorides K_3NbOF_6 and K_3TaOF_6. *Rev.Chim.Miner.* **24(5)**, 583.

FOX, D.L., SCOTT, J.F. & BRIDENBAUGH, P.M. (1976). Soft modes in ferroelastic LaP_5O_{14} and NdP_5O_{14}. *Sol.State Comm.* **18**, 111.

GENE, V.V. & SHUL'GA, V.I. (1982). Appearance of an incommensurate phase in proper ferroelastic phase transitions from the tetragonal system. *Sov. Phys.* **24(5)**, 902.

GOMEZ-CUEVAS, A., ECHARRIE, A.L., TELLO, M.J. & RIVEZ-OLAELA, J. (1981). Critical behaviour of the ferroelastic-incommensurate $(N(CH_3)_4)_2CuCl_4$. *Ferroelectrics* **36(1-4)**, 384.

GOMEZ-CUEVAS, A., LAUNAY, J.C., PEREZ MATO, J.M., FERNANDEZ, J., ECHARRIE, A.L. & TELLO, M.J. (1984). Experimental and group theoretical study of the structural phase transition in $RbMnCl_3$ crystals. *Ferroelectrics* **55(1-4)**, 789

GOMEZ-CUEVAS, A., TELLO, M.J., FERNANDEZ, J., ECHARRIE, A.L., HERNEROS, J. & COUZI, M. (1983). Thermal, optical and spectroscopic investigations in ferroelastic incommensurate $(N(CH_3)_4)_2CuCl_4$ crystal. *J.Phys.C* **16(3)**, 473

GOREV, M.V., MELNIKOVA, S.V. & FLEROV, I.N. (1987). Thermophysical and optical studies of the ferroelastic Rb_2CdCl_4. *Sov.Phys.* **29(7)**, 1199.

GORIN, YU.F., BABUSHKIN, A.N., KOBELEV, L.YA.& SAVEL'KAEV, A.S. (1985). Ferroelastic properties of proustite at low temperatures. *JETP Lett.* **41**, 521.

GREKOV, A.A., NITSCHE, R., KOSONOGOV, N.A., ROGAIN, E.D., RODIN, A.I. & FRIDKIN, V.M.(1981). Optically induced switching in ferroelastic Sb_5O_7I. *Sov.Phys.* **23(4)**, 689.

GRIDNEV, S.A. & SHUVALOV, L.A. (1983). The influence of real structure on switching processes and peculiarities of mechanical relaxation in proper ferroelastics $KH_3(SeO_3)_2$ and $KD_3(SeO_3)_2$. *Ferroelectrics* **48(1-3)**, 169.

GRIDNEV, S.A., DARINSKII, B.M. & PRASOLOV, B.N. (1983). Internal friction mechanisms in improper ferroelectric-ferroelastic $Gd_2(MoO_4)_3$. *Ferroelectrics* **48(1-3)**, 169.

GRIDNEV, S.A., KUDRYASH, V.I. & SHUVALOV, L.A. (1979a). Mechanical hysteresis loops in $KH_3(SeO_3)_2$ crystals. *Bull.Acad.Sci.USSR Phys.Ser.* **43(8)**, 145.

GRIDNEV, S.A, DARINSKII, B.M., PRASOLOV, B.N., SHUVALOV, L.A. & FEDOSYUK, R.M. (1979b). Several thermodynamic characteristics of $KH_3(SeO_3)_2$ and $KD_3(SeO_3)_2$ crystals from ultralow-frequency measurements of the shear modulus. *Bull. Acad. Sci. USSR, Phys. Ser.* **43(8)**, 140.

GRIDNEV, S.A., KUDRYISH, V.I. & PRASOLOV, B.N. (1980). Anelastic and elastic infra-low frequency properties of ferroelastic crystals $K(D_xH_{1-x})_3(SeO_3)_2$. *Ferroelectrics* **26(1-4)**, 669.

GRIDNEV, S.A., SHUVALOV, L.A. & BONDARENKO, V.V. (1985). Ferroelastic phase intermediate between alpha and beta phases in a $NaH_3(SeO_3)_2$ crystal. *Sov. Phys.* **27**, 283.

GUMENYUK, A.F., OLEINIK, O.I. & OMELYNANENKO, V.A. (1987). Effect of ferroelastic phase transition on the spectra of edge absorption and Raman scattering in barium sodium niobate single crystals. *Opt.Spectrosc.* **62(2)**, 221.

GUNTON, D.J. & SAUNDERS, G.A. (1974). Elastic behaviour of In-Tl alloys in the vicinity of the martensite transition. *Sol.State Comm..* 14 865.

GURSKAS, A.A., KOKSHANEV, V.B.& SYRKIN, E.S. (1987). Dynamical character of impurity influence on phase transitions in layered ferroelastics. *Ferroelectrics* **75**, 483.

GURSKAS, A.A., ZVYAGIN, A.I. & PELIKH, L.N. (1983). Ferroelastic phase transitions in $CsR(MoO_4)_2$-$CsR(WO_4)$ with R=Dy, Ho, Er, Tm, Yb and Lu. *Ferroelectrics* **48(1-3)**, 81.

GURSKAS, A.A., KOKSHANEV, V.B. & SYRKIN, E.S. (1985). Phase transitions in layered ferroelastics doped with light impurities. *Sov. Phys.* **27**, 1250.

HAGEN, R.M. (1982). Bismuth vanadate: a high-pressure, high-temperature crystallographic study of the ferroelastic-paraelastic transition. *Science* **216(4549)**, 991.

HAMARDA, S., GURSKAS, A.A. & POPOV, V. (1989). Influence of substitution impurity on the temperature of phase change of the stratified $CsDy(MoO_4)_2$. *C. R. Acad. Sci. II, Mec. Phys. Chim. Sci. Univers. Sci. Terre* **308(17)**, 1529.

HARLEY, R.T., LOWDE, R.D., SAUNDERS, G.A.,SCHERM, M.& UNDERHILL C.I.(1978).in *Proc. Int. Conf. on Lattice dynamics..* Paris Ed. Balkanski (Flammarion,Paris.).

HARRIS, M.J.,SALJE, E., GUTTLER, B.K., & CARPENTER,M.A. (1989). Determination of the degree of Al,Si order Q_{od} in kinetically disordered albite using hard mode infrared spectroscopy.*Phys.Chem.Min.*16 649.

HATTORI, A., WADA, M., SAWADA, A. & ISHIBASHI, T. (1980). Raman scattering study of ferroelastic phase transition in $C_6H_5NH_3Br$ crystal. *J. Phys. Soc. Japan.* 49(2), 624.

HAUSSÜHL, S.(1973). Anomalous thermoelastic behaviour of cubic potassium cyanide.*Sol.State Comm.* 13 147.

HAUSSÜHL, S., ECKSTEIN, J., RECKER, K. & WALLRAFEN, F. (1977). Cubic sodium cyanide, an other crystal with KCN-type anomalous thermoelastic behaviour. *Acta Cryst. A* 33, 847.

HAUSSÜHL, S. (1984). Ferroelastic phase transition and anomalous elastic and thermal properties of betaine borate $(CH_3)_3NCH_2COO.H_3BO_3$. *Solid State Commun.* 50(1), 63.

HEBBACHE, M., ERRANDONEA, G. & SCHNECK, J. (1984). Brillouin scattering study of the ferroelastic-incommensurate transition in $KLiSO_4$ barium sodium niobate. *Ferroelectrics* 55(1-4), 707.

HEILMAN, I.U.,ELLENSON W.D.& ECKERT J.(1979). Softening of transverse elastic modes at the structural phase transition of s-triazine, $C_3N_3H_3$. *J.Phys.C.Solid State Phys.* L185.

HEMPEL, H., MAACK, H. & SORGE, G. (1988). Ferroelastic phase transition in $LiRb_4H(SO_4)_3.H_2SO_4$. Landau theory and experimental results. *Phys.Status Solidi A* 110(2), 459.

HIKITA, T., SEKIGUCHI, H. & IKEDA, T. (1977). Phase transitions in new langebeinite-type crystals. *J.Phys. Soc. Japan* 43, 1327.

HIROTSU, S. (1983). Birefringence study of the Potts model phase transition in NaN_3. *J. Phys. C* 16(31), L1103.

HRABANSKI, R. (1988). EPR study of the ferroelastic phase transition in Mn^{2+}-doped $[Co(H_2O)_6]GeF_6$ single crystals. *Ferroelectrics* 80, 663.

HUANG, S.-J., & Yu, J.-T. (1987). Ferroelastic transition of $LiCsSO_4$ studied by the EPR of $.NH_3^+$. *Solid State Commun.* 63, 745.

ISHIBASHI, Y., HARA, K. & SAWADA, A. (1988). The ferroelastic transition in some scheelite-type crystals. *Physica B & C* 150 B+C(1-2), 258.

IVANOV, N.R. (1979). Thermal expansion and spontaneous strain in the ferroelastics $KH_3(SeO_3)_2$, $KD_3(SeO_3)_2$ and $K(H_{0.2}D_{0.8})_3(SeO_3)_2$. *Bull. Acad. Sci. USSR, Phys. Ser.* 43(8), 134.

IVLIEV, M.P. & SAKHNENKO, V.P. (1979). Theory of ferroelastic phase transitions in KCN. *Bull. Acad. Sci. USSR, Phys. Ser.* 43(8), 48.

IVLIEV, M.P. & SAKHNENKO, V.P. (1986). Possibility of existence of a quadropole ferroelastic phase in $KTaO_3:Li$. *Sov. Phys.* 28, 356.

JANG, M.S. & LEE, J.H. (1983). Ferroelastic phase transition in $C_9H_{18}NO$ single crystals.*J. Korean Phys. Soc.* 16(1), 59.

JEX, H., MAETZ, J., & MULLNER, M.(1980). Cubic-to-tetragonal phase transition in $RbCaF_3$ investigated by diffraction experiments with neutrons, X-rays and gamma rays from a Mossbauer source.*Phys.Rev. B* 21 1209

JORGENSEN, J.D., WORLTON, T.G. & JAMIESON, J.C. (1978). Pressure induced strain transition in NiF$_2$. *Phys. Rev. B*. **17**, 2212.

KAPLAN, M.D. (1988). Low temperature reorientation structural transitions in Jahn-Teller DyVO$_4$ ferroelastic induced by external magnetic field. *Fiz. Nizk. Temp.* **14(1)**, 62 (In Russian).

KAPLYANSKII, A.A., MARKOV, YU.F. & BARTA, C. (1979). Mercury (I) halides - a new group of pure improper ferroelastics. *Bull. Acad. Sci. USSR, Phys. Ser.* **43(8)**, 77.

KASSEM, M.E.M. & MROZ, B. (1983). Elastic anomaly in LiKSO$_4$ crystals. *Acta Phys. Pol. A* **A63(4)**, 449.

KIAT, J.M., CALVARIN, G., GARNIER, P. & GREGOIRE, P. (1985). Ferroelastic phase transitions in lead orthophosphovanadates Pb$_3$P$_{2x}$V$_{2-2x}$O$_8$ *Japan. J. Appl. Phys. Suppl.* **24**, 690.

KIAT, J.M., GARNIER, P., CALVARIN, G. & WEIGEL, D. (1984). Phase transitions and acousto-optic properties of lead orthophosphovanadates. *Ferroelectrics* **55(1-4)**, 803.

KIHARA, K., (1990). An X-ray study of the temperature dependenceof the quartz structure. *Eur. J. Mineral.* **2**, 63.

KINO, Y., LÜTHI, B..& MULLEN, M.E. (1973). Elastic properties and cooperative Jahn-Teller effect in nickel chromite. *Sol. State. Comm.* **12**, 275.

KIOSSE, G.A., DUDNIK, E.F. & SUSHKO, S.A. (1981). Ferroelastic properties of lead orthoarsenate Pb$_3$(AsO$_4$)$_2$ *Bull. Acad. Sci. USSR, Phys. Ser.* **45**, 139.

KIOSSE, G.A., DUDNIK, E.F., SUSHKO, S.A. & KOLESOV, I.S. (1982). X-ray and optical investigations of ferroelastic single crystals of Pb$_8$V$_2$O$_{13}$ and Pb$_8$P$_2$O$_{13}$. *Sov. Phys. - Crystallogr.* **27(6)**, 713.

KISHIMOTO, T., OSAKA, T., KOMUKAE, M. & MAKITA, Y. (1987). Ferroelastic phase transition in (NH$_4$)$_3$H(SeO$_4$)$_2$. *J.Phys.Soc Japan* **56(6)**, 2070.

KJEMS,J .K., SHIRANE, G., BIRGENEAU, R.J. & VANUITERT, L .G. (1973). Quadrupole exciton-phonon dynamics at 151K phase transition in PrAlO$_3$. *Phys Rev. Lett.* **31** 1300.

KLEEMAN, W., SCHAFER, F.J. & NOUET, J. (1982). Structural phase transition in ferroelastic RbAlF$_3$ II. Linear birefringence investigations.*J. Phys. C* **15(2)**, 197.

KLEVTSOVA, R.F., SOLODOVNIKOV, S.F. & KLEVTSOV, P.V. (1986). The structural changes in the phase transition in the ferroelastic K$_4$Zn(MoO$_4$)$_3$. *Bull. Acad. Sci. USSR, Phys. Ser.* **50**, 137.

KNORR, K., LOIDL, A. & KJERM, J.K. (1986). Ferroelastic transition in KBr:KCN studied by neutrons, X-rays and ultrasonics. *Physica B+C* **136B+C**, 311.

KNORR, K., LOIDL, A. & KJERM, J.K. (1985). Continuous ferroelastic phase transition of a KBr:KCN mixed crystal. *Phys. Rev. Lett* **55**, 2445.

KOLPAKOVA, N.N., MARGRAF, R., & PIETRASZKO, A. (1987). A new ferroelastic: cadmium pyroniobate Cd$_2$Nb$_2$O$_7$. *Sov. Phys.* **29(9)**, 1520.

KRAINYUK, G.G., NOSENKO, A.E. & OTKO, A.I. (1983a). Techniques for investigation of ferroelastic stress-strain curve peculiarities related to domain structure rearrangement. *Ferroelectrics* **48(1-3)**, 175.

KRAINYUK, G.G., OTKO, A.I. & NOSENKO, A.E. (1983b). Switching of ferroelastic domains in $KFe(MoO_4)_2$ and related crystals. *Bull. Acad. Sci. USSR, Phys. Ser.* **47(4)**, 132.

KRAJEWSKI, T., BRECZEWSKI, T., PISKUNOWICZ, P. & MROZ, B. (1985). High temperature ferroelastic phase in $LiKSO_4$ crystals. *Ferroelectr. Lett. Sect.* **4**, 95.

KROUPA, J., FOUSEK, J., FOUSKEIVA, A., PETZELT, J., PAVEL, M., KAMBA, S., FUITH, A. & WARHANEK, H. (1988). New data on the phase transition in KSCN. *Ferroelectrics* **87**, 121.

KUKUEVA, L.L., IVANOVA, L.A. & VENEVTSEV, YU.N. (1984). Ferroelastics with the fergusonite type structure. *Ferroelectrics* **55(1-4)**, 797.

KULIKOV, N.I. & TUGUSKEV, V.V. (1981). Electronic theory of the structural phase transition in the dihydrides of the IVB transition metals. *Sov. Phys.* **23(9)**, 1631.

LAJZEROWICZ, J., LEGRAND, J.F. & JOFFRIN, C. (1980). Ferroelastic and ferroelectric phase transitions in a molecular crystal: tenane. I. Neutron and Brillouin scattering studies of the acoustic and pseudo-spin coupled modes. *J. de Phys.* **41(11)**, 1375.

LANZE DE DUSSEL, H., DE WAINER, L.S., DE BENYICAR, M.A.R. & SA, C. (1981). Synthetic troegenite, phase II: ferroelastic character and possible point groups. *Ferroelectrics* **39(1-4)**, 1067.

LIM,A.R., CHON S.H.& JUNG M.S.(1988). Domain structure and relaxation time of ^{51}V NMR in ferroelastic $BiVO_4$. *New Phys. (Korean Phys. Soc.)* **28(5)**, 611.

LUTHI, B., MULLEN, M.E., ANDRES, K.BUCHER, E &MAITAJ. P. (1973)..Experimental investigation of cooperative Jahn-Teller effect in TmCd.*Phys.Rev.B* **8** 2639.

LUTY, T. & VAN DER AVOIRD, A. (1983). Role of translational-rotational coupling in lattice dynamics and ferroelastic phase transitions;the s-triazine crystal.*Chem. Phys.* **83 (1-2)**,133.

MACKENZIE, G.A. & PAWLEY, G.S., (1979). Neutron scattering study of DCN. *J. Phys.* C **12**, 2717.

MAKITA, Y., SAKURAI, F., OSAKA, T.& TATSUZAKI, I.(1977). Study of phase transition in $KH_3(SeO_3)_2$.I -Acoustic phonon softening.*J.Phys.Soc.Jap.***42** 518.

MANOLIKAS, C. (1981). Electron microscopic study of the commensurate-incommensurate phase transition in $Ba_2NaNb_5O_{15}$ and $Sr_2KNb_5O_{15}$. *Phys.Status Solidi a* **68(2)**, 653.

MANOLIKAS, C. & AMELINCKX, S.(1980). Phase transition in ferroelastic lead orthovanadate as observed by means of electron microscopy and electron diffraction. I. Static observations. *Phys.Status Solidi A* **60(2)**, 607.

MARTIN, B., PASTOR, J.M., RULL, F. & DE SAJA, J.A. (1982). A study on the structural phase transitions of ferroelastic $[N(CH_3)_4]_2CuCl_4$ from micro hardness measurements. *Solid State Commun.* **44(7)**, 1047.

MATSUSHITA, E. & MATSUBARA, T. (1981). Theory of ferroelastic phase transition in squaric acid. *Ferroelectrics* **39(1-4)**, 1095.

MELCHER, R.L.&SCOTT, B.A(1972). Soft acoustic mode at cooperative Jahn-Teller phase transition in DyO_4.*Phys Rev.Lett..* **28** 607

MELCHER, R.L., PYTTE E.& SCOTT, B.A. (1973) Phonon instabilities in TMVO$_4$.*Phys.Rev. Lett.***31** 307.

MIDORIKAWA, M., TSUNODA, K. & ISHIBASHI, Y. (1980). Physical properties related to phase transitions in Na$_2$CO$_3$ crystals. *J.Phys.Soc.Japan.* **49(1)**, 242.

MOGUS-MILANKOVIC, A., RAVEZ, J., CHAMINADE, J.P., HAGENMULLER, P. (1985). Ferroelastic properties of TF$_3$ compounds (T=Ti, V, Cr, Fe, Ga). *Mater. Res. Bull.* **20**, 9.

MORAIS, P.C., RIBEIRO, G.M. & CHAVES, A.S. (1984). EPR study of the ferroelastic phase transition in CsLiSO$_4$. *Solid State Commun.* **52**, 291.

MOORE, D.R., TEKIPPE, V.J., RAMDAS, A.K.& TOLEDANO, J.C.(1983). Solid-to-solid phase transition in crystalline benzil-a Raman scattering study.*Phys Rev B* **27** 12 7676.

MORAN, T.J., THOMAS, R.L., LEVY, P.M.& CHEN, H.H.(1973). Elastic properties of DySb near magnetic and structural phase transitions.*Phys.Rev.* **B 7** 3238.

MROZ, B., KIEFTE, H. & COULTER, M.J. (1988). Identification of the ferroelastic phase transition in LiRb$_5$(SO$_4$)$_3$><$_{1.5}$H$_2$SO$_4$ by Brillouin spectroscopy. *Ferroelectrics* **82**, 105.

MROZ, B., TUSZYNSKI, J.A., KIEFTE, H. & CLOUTER, M.J. (1989). On the ferroelastic phase transition of LiNH$_4$SO$_4$: a Brillouin scattering study and theoretical modelling. *J.Phys.C.* **1(4)**, 783.

MNUSHKINA, I.E. & DUDNIK, E.F. (1982). Domain structure in single crystals of bismuth vanadate. *Sov.Phys. Crystallogr.* **27(4)**, 485.

NAKAYAMA, H., ISHII, K., CHIJUWA, E., WADA, M. & SAWADA, A. (1985). Phase transition and ferroelasticity of the phenothiazine crystal. *Solid State Commun.* **55**, 59.

OHNO, E. (1987). Theory of ferroelastic phase transition in NaCN, KCN and RbCN. *J. Phys. Soc. Japan* **56(12)**, 4414.

OSAKA, T., MAKITA, Y. & GESI, K. (1977). Ferroelectricity of deuterated triammonium deuterium disulphate and isotope effect on ferroelectric activity. *J. Phys. Soc. Japan* **43**, 933.

OTKO, A.I., KRAINYUK, G.G., STOLPAKOVA, T.M. & DUDNIK, E.F. (1984).. Domain switching and crystallographic features of monoclinic ferroelastic phases of some double molybdates and tungstates. *Bull. Acad. Sci. USSR, Phys. Ser.* **48**, 73.

OTKO, A.I., NESTERENKO, N.M. & ZUYAGIN, A.I. (1979). Ferroelastic phase transitions in trigonal dimolybdates and ditungstenates. *Bull. Acad. Sci. USSR, Phys. Ser.* **43(8)**, 108.

OTKO, A.I., NESTERENKO, N.M., KRAINYUK, G.G. & NOSENKO, A.E. (1983). KFe(MoO$_4$)$_2$ and KIn(WO$_4$)$_2$ ferroelastic domain structures. *Ferroelectrics* **48(1-3)**, 143.

OZEKI, H. & SAWADA, A.(1982). Acoustic softening in ferroelastic LiCsSO$_4$ crystal. *J.Phys.Soc.Japan* **51(7)**, 2047.

PAKULSKI, G., MROZ, B. & KRAJEWSKI, T. (1983). Ferroelastic properties of LiCsSO$_4$ crystals. *Ferroelectrics* **48(4)**, 259.

PAQUET, D. & LEROUX-HAGON, P. (1979). Electron correlations and electron-lattice interactions in the metal-insulator, ferroelastic transition in VO_2: a thermodynamical study. *Phys.Rev.B* **22(11)**, 5284.

PARFENOV, O.E. & CHERNYSHOV, A.A. (1988). Emission of sound from metallic ferroelastics exhibiting an FCC-FCT transition. *Sov.Phys.***30(9)**, 1611.

PEERCY,P.S. & FRITZ,I.J.(1974) Pressure induced phase transition in paratellurite(TeO_2).*Phys. Rev. Lett.*. **32** 466.

PELIKH, L.N. (1983a). Effect of uniaxial compression on the ferroelastic phase transitions in layered $KSc(MoO_4)_2$. *Sov. Phys.* **24(10)**, 1782.

PELIKH, L.N. (1983b). Phase transitions in laminated $K_{1-x}NaxSc(MoO_4)_2$ ferroelastics. *Ukr. Fiz. Zh.* **28(6)**, 939. (In Russian)

PLAKIDA, N.M. & SALEJDA, W. (1988a). The improper ferroelastic phase transition in superionic $Rb_3H(SeO_4)_2$ crystals. *Phys.Status Solidi B* **148(2)**, 473.

PLAKIDA, N.M. & SALEJDA, W. (1988b). The phenomenonlogical theory of the improper ferroelastic phase transition in superionic $Rb_3H(SeO_4)_2$ crystals. *Ferroelectrics* **81**, 1171.

PLAKIDA, N. & SHAKHMATOV, V.S. (1987). The non-inherent ferroelastic transition in $CsHSO_4$-type superionic crystals. *Bull. Acad. Sci. USSR, Phys. Ser.* **51(12)**, 24.

PLESKO, S., DVORAK, V., KIND, R. & TREINDL, A. (1981). Successive phase transitions in pseudohexagonal Cs_2HgBr_4. *Ferroelectrics* **36(1-4)**, 331.

POUGET, J. (1989). Nonlinear dynamics for proper ferroelastic transitions for which the Landau condition is violated. *Phase Transitions* **14(1-4)**, 251.

PREWITT, C.T. (1970). X-ray diffraction study of $Gd_2(MoO_4)_3$. *Sol. State. Comm.* **8**, 2037.

PRINGLE, G.E. & NOAKES, D.E., (1968). Crystal structures of lithium, sodium and strontium azides. *Acta Cryst.* **B24**, 262.

RAE, A.I.M. (1972). Lattice dynamics of HCN. *J. Phys.* C **5**, 3309.

RAE, A.I.M.(1982). The structural phase change in s-triazine:the quasi-harmonic approximation.*J.Phys.C*.**15** 1883.

RAICH, J.C.& BERNSTEIN, E.R. (1982). Comment on the quasi-harmonic treatment of the structural phase change in s-triazine [and reply]. *J. Phys. C* **15(10)**, L283.

RAICH, J.C., YOSHIKARA, A. & BERNSTEIN, E.R. (1982). An equation of motion approach to ferroelasticity in sym-triazine. *Mol. Phys.* **45(1)**, 197.

RAICH, J.C., YASUDA, H. & BERNSTEIN, E.R. (1983). Mean field approach to ferroelastic transitions in molecular crystals. *J. Chem. Phys.* **78(10)**, 6209.

RAVEZ,J., ELAATMANI, M.& CHAMINADE, J.P.(1979). Preparation, etudes optiques et dielectriques de monocristaux de $Na_5W_3O_9F_5$.*Sol.State Comm.*. **32**, 749

RAVEZ, J. & MOGUS-MILANKOVIC, A. (1985). The ReO_3-related structure ferroelastic fluorides. *Japan. J. Appl. Phys. Suppl.* **24**, 687.

RAVEZ, J., MOGUS-MILANKOVIC, A., CHAMINADE, J.P. & HAGENMULLER, P. (1984). Ferroelastic properties of AlF_3. *Mater. Res. Bull.* **19**, 1311.

RAVEZ, J., ABRAHAMS, S.C., ZYONTZ, L.E. & CHAMINADE, J.P. (1983)..
Mechanical de-twinning of $Na_5W_3O_9F_5$ ferroelastic crystals. *Solid State Chemistry Conf. Proc.*, 745.

REDFERN, S.A.T. & SALJE, E. (1988). Spontaneous strain and the ferroelastic phase transition in As_2O_5. *J. Phys. C* **21(2)**, 277.

REESE.R.L., FRITZ, I.J., &CUMMINS, H.Z.(1973). Light scattering studies of soft ferroelastic and acoustic modes of KD_2PO_4.*Phys,Rev. B* 4165.

REHWALD, W., VONLANTHER, A., REHABER, E. & PRETTL, W. (1980). The elastic behaviour of ferroelastic Sb_5O_7I polytype 2MC studied by ultrasonic experiments and Brillouin scattering. *Z. Phys. B* **39(4)**, 299.

REHWALD, W. (1968). Lattice softening and stiffening of single crystal Niobium Stannide at low temperatures. *Phys. Lett. A* **27**, 287.

RIDOU, C., ROUSSEAU, M. & NOUET, J. (1980). Phenomenological description of the first order improper ferroelastic phase transition in $RbCaF_3$ at 194K. *Ferroelectrics* **26(1-4)**, 685.

ROCQUET, P. & COUZI, M. (1985). Structural phase transition in chiolite $Na_5Al_3F_{14}$. II. A model for the pseudo-proper ferroelastic transition. *J. Phys. C* **18**, 6571.

ROUCAU, C., AYROLES, R. & JORNES, J. (1983). Electron microscopy study of the domain wall structure of ferroelastic phase $Pb_3(PO_4)_2$. *J. Phys.* **44(2)**, 141. (In French).

ROUSSEAU, M., BULOU, A.,RIDOU, C., & HEWAT, A.W.(1980). A new model for phase transitions due to octahedra rotations in a perovskite structure. *Ferroelectrics* **25** 447.

ROWE, J.M., RUSH, J.J., VAGELATOS, N., PRICE, D.L., HINKS, D.G. & SUSSMAN, S. (1975). Crystal dynamics of KCN and NaCN in disordered cubic phase. *J. Chem. Phys.* **62**, 4551.

ROWE,J .M., RUSH, J.J. & LUTHI, F.(1984). Crystal structure of rubidium cyanide at 4K determined by neutron powder diffraction. *Phys. Rev. B* **29** 2168.

SADOVSKAYA, L.J.A., DUDNIK, E.F., SCHERBINA, S.A. & GRZHEGORZHEVSKII, O.A. (1983). Ferroelastic properties of ditellurites. *Ferroelectrics* **48(1-3)**, 109.

SAINT-GREGOIRE, P., ALMAIRAL, R., FREUND, A. & GESTANE, J.Y. (1986). Barium manganese fluoride $BaMnF_4$ as an improper ferroelastic. *Ferroelectrics* **67**, 15.

SAKHNENKO, V.P. & TIMONIN, P.N. (1983). Fluctuation anomalies of the elastic and dielectric properties of improper ferroelectric ferroelastics belonging to the gadolinium molybdate family. *Sov. Phys.* **24(12)**, 2055.

SALJE, E., KUSCHOLKE, B.WRUCK, B.&KROLL H. (1985).Thermodynamics of sodium feldspar 2-experimental results andnumerical calculations. *Phys.Chem.Min.* **12** 132.

SALJE, E. & WRUCK, B. (1983). Specific-heat measurements and critical exponents of the ferroelastic phase transition in $Pb_3(PO_4)_2$ and $Pb_3(P_{1-x}As_xO_4)_2$. *Phys.Rev. B* **28(11)**, 6510.

SALJE, E., BISMAYER, U. & JANSEN, M. (1987). Temperature evolution of the ferroelastic order parameter of As2O5 as determined from optical birefringence. *J. Phys. C* **20**, 3613.

SALJE, E. & DEVARAJAN, V. (1981). Potts model and phase transition in lead phosphate $Pb_3(PO_4)_2$. *J. Phys. C* **14(33)**, L1029.

SALJE, E., DEVARAJAN, V., BISMAYER, U. & GUIMARAES, D.M.C. (1983). Phase transitions in $Pb_3(P_{1-x}As_xO_4)_2$: influence of the central peak and flip mode on the Raman scattering of hard modes. *J. Phys. C.* **16(27)**, 5233.

SALJE, E. & PARLINSKI, K., (1991). Microstructures in the high T_c superconductors. *Supercond. Science and Technology* **4**, 93.

SALJE, E., RIDGWELL, A., GÜTTLER, B., WRUCK, B., DOVE, M.T. & DOLINO, G., (1991). On the displacive character of the phase transition in quartz: a hard-mode spectroscopic study. *J. Phys. Cond. Matt.* **3**, 1.

SANDERCOCK J.R., PALMER S.B., ELLIOTT R.J. HAYES, W.,SMITH S.R.P. & YOUNG A.P.(1972). Brillouin scattering ultrasonic and theoretical studies of acoustic anomalies in crystals showing Jahn-Teller phase transitions.*J.Phys.C* **5** 3126.

SANDLER, Yu.M. & SERDOBOL'SKAJA, O.Yu. (1983). Nonlinear elastic waves in ferroelastics. *Ferroelectrics* **48(1-3)**, 49.

SANDLER,Yu.M., ZAITSEVA M.P., SYSOEV A.I.& KOKORIN (1978). Yu.I.Non-linear electro-mechanical properties of ferroelectric $NaND_4SeO_42D_2O$. *Ferroelectrics* **20** 531 .

SATO, M., SOEJIMA, Y., OHAMA, N., OKAZAKI, A., SCHEEL, H.J. & MULLER, K.A. (1985). The lattice constant vs. temperature relation around the 105K transition in a flux-grown $SrTiO_3$ crystal. Phase Trans. **5**, 207.

SAUNDERS, G.A., COMINS, J.D., MACDONALD, J.E. & SAUNDERS, E.A. (1986). Thermodynamics of a ferroelastic phase transition. *Phys. Rev. B* **34**, 2064.

SAWADA, A., UDAGAWA, M. & NAKAMURA, T. (1977). Proper ferroelastic transition in piezoelectric lithium ammonium tartrate.*Phys Rev.Lett.* 39 829.

SAWADA, A., HATTORI, A. & ISHIBASHI, T. (1981). Light scattering study of ferroelastic transition in $C_6H_5NH_3Br$ crystals. *Ferroelectrics* **39(1-4)**, 1147.

SAWADA, A., HATTORI, A. & ISHIBASHI, T. (1980a). Acoustic softening in ferroelastic $C_6H_5NH_3Br$ crystal. *J.Phys.Soc.Japan.* **49(1)**, 423.

SAWADA, A., SUGIYAMA, J., WADA, M. & ISHIBASHI, Y. (1980b). Evidence of incommensurate-ferroelastic (commensurate) phase transition in $[N(CH_3)_4]_2CuCl_4$ crystal. *J.Phys.Soc.Japan* **48(5)**, 1773.

SAWADA, A., SUGIYAMA, J. & ISHIBASHI, Y. (1981). Acoustic softening in incommensurate-ferroelastic transition in $\{N(CH_3)_4\}2CuCl_4$ crystal. *Ferroelectrics* **36(1-4)**:385.

SAWADA, A. & NAKAMURA, T. (1982). Intrinsic ferroelastic transition from a piezoelectric paraelastic phase. *Ferroelectrics* **40(3-4)**, 141.

SCHMAHL, W.W., PUTNIS, A., SALJE, E., FREEMAN, P., GRAEME-BARBER, A., JONES, R., SINGH, K.K., BLUNT J., EDWARDS, P.P., LORAM, J. & MIRZA, K. (1989). Twin formation and structural modulaations in orthorhombic and tetragonal $YBa_2(Cu_{1-x}Co_x)_3O_{7-\delta}$. *Phil. Mag. Lett.* **60**, 241.

SCHMAHL, W.W. & REDFERN , S.A.T. (1988). An X-ray study of coupling between acoustic and optic modes of the ferroelastic phase transition in As_2O_5. *J. Phys. C* **21(20)**, 3719.

SCHNECK, J. & DENOYER, F. (1981). Incommensurate phases in barium sodium niobate. *Phys. Rev. B* **23(1)**, 383.

SCHNECK, J., TOLEDANO, J.C., JOFFRIN, C., AUBREE, J., JOUKOFF, J. & GABELOTAUD, A. (1982). Neutron scattering study of the tetragonal-to-incommensurate ferroelastic transition in barium sodium niobate. *Phys. Rev. B* **25(3)**, 1766.

SCHNEIDER, V.E., VLASOVA, A.A. & IVANOV, N.R. (1987). Microscopic theory of order-disorder phase transitions in pure ferroelastic crystals. *Ferroelectrics* **75**, 419.

SHAPIRO, S.M., AXE, J.D. & SHIRANE, G. (1972). Critical neutron scattering in $SrTiO_3$ and $KMnF_3$. *Phys. Rev. B*. **6(11)**, 4332.

SHCHEDRINA, N.V. (1980). Thermal anomalies near the phase transition point in $RbMnCl_3$-type ferroelastics. *Sov. Phys.* **22(9)**, 1490.

SIVARDIERE, J.(1972). Theory of phase transitions in rare earth vanadates. *Phys.Review B* **6** 4284

SKELTON, E.F., FELDMAN, J.L., LIU, C.Y. & SPAIN, I.L. (1976). Study of pressure induced phase transition in paratellurite(Teo-2).*Phys.Rev. B* **13** 2605.

SKOROBOGATOVA, I.V. & SAVCHENKO, E.M. (1983). Optic investigation of the ferroelastic transition in $CsHo(MoO_4)_2$. *Ferroelectrics* **48(1-3)**, 87.

SKOROBOGATOVA, I.V., SAVCHENKO, E.M. & ZVYAGIN, A.I. (1983). Features of the pT diagram at the ferroelastic transition in $CsDy(MoO_4)_2$. *Bull.Acad.Sci.USSR, Phys. Ser.* **47(3)**, 69.

SKOROBOGATOVA, I.V.,SAVCHENKO, E.M.& VOLOSHIN, V.A.(1986). Effects of hydrostatric pressure on the phase transition in $CsDy(MoO_4)_2$ *Bull.Acad. Sci. USSR Phys. Ser.* **50** 164.

SKOROBOGOTOVA, I.V.,SAVCHENKO, E.M. & ZVYAGIN, A.I.(1983). Features of the pT diagram at the ferroelastic transition in $CsDy(MoO_4)_2$ *Bull.Acad.Sci.USSR Phys. Series* **47** 69.

SMITH,J.H. & RAE, A.I.M. (1978a). Structural phase change in s-triazine.1-Crystal structure of low temperature phase.*J.Phys.C.Solid State* **11** 1761.

SMITH, J.H.& RAE, A.I.M. (1978b). Structural phase change in s-triazine.2-Specific heat measurements.*J.Phys.C.Solid state* **11** 1771.

SMIRNOV, P.S., STRUKOV, B.A., GORELIK, V.S. & DUDNIK, E.F. (1979). Raman scattering of light by low-frequency oscillations in the improper ferroelastic $Pb_3(PO_4)_2$. *Bull.Acad.Sci.USSR, Phys. Ser.* **43(8)**, 104.

SMOLENSKII, G.A., SINII, I.G., KUZ'MINOV, E.G. & DUDNIK, E.F. (1979a). Optical phonons and soft modes in the ferroelastic $KFe(MoO_4)_2$. *Bull.Acad.Sci.USSR, Phys. Ser.* **43(8)**, 85.

SMOLENSKII, G.A., SINII, I.G., ARNDT, KH., PROKHOROVA, S.D., KUZ'MINOV, E.G., MIKVABIYA, D., & KOLPAKOVA, N.N. (1979b). Light scattering in trisarcosine calcium chloride (TSCC). *Bull.Acad.Sci.USSR, Phys. Ser.* **43(8)**. 99.

SMOLENSKII, G.A., PROKHOROVA, S.D., SING, I.G., FOUSKOVA, A., KONAN, G. & DUDNIK, E.F. (1980). Phase transition in ferroelastic $KFe(MoO_4)_2$. *Ferroelectrics* **26(1-4)**, 677.

SOL'TSAS, R.KH. & SCHNEIDER, V.E. (1983). Statistical models of the phase transition in $KH_3(SeO_3)_2$. *Sov. Phys. - Crystallogr.* **28(2)**, 123.

SORGE, G., ALMEIDA, A., SHUVALOV, L.A. & FEDOSYUK, R.M. (1981). Influence of the mechanical stress component T5 on the elastic behaviour of $KH_3(SeO_3)_2/KD_3(SeO_3)_2$ crystals in the vicinity of their ferroelastic phase transitions. *Phys. Status Solidi a* **68(1)**, 245.

SORGE, G., BEUGI, H., ALMEIDA, A. & SHUVALOV, L.A. (1982a). Elastic nonlinearity of $KH_3(SeO_3)_2$ and $KD_3(SeO_3)_2$ crystals. *Phys. Status. Solidi a* **73(1)**, K63.

SORGE, G., STRAUBE, U. & ALMEIDA, A. (1982b). Investigations of ferroelastic phase transition by ultrasonic methods. *Acta Phys. Slovaka* **32(1)**, 55.

STEVENS, E.D. & HOPE, H. (1977). Study of electron-density distribution in sodium azide NaN_3. *Acta Cryst.* **A33**, 723.

SUGIYAMA, J., WADA, M., SAWADA, A. & ISHIBASHI, Y. (1980). Successive phase transitions in $\{N(CH_3)_4\}_2CuCl_4$. *J. Phys. Soc. Japan.* **49(4)**, 1405.

SUUMI, K.A., KOYIMA, S. & NAKAMURA, T. (1980). Effect of the hydrostatic pressure on the ferroelastic NdP_5O_{14}. *J.Phys.Soc.Japan.* **48(4)**, 1298.

SUZUKI, I. & ISHIBASHI, Y. (1987). Phenomenological considerations of the electric field induced transitions in improper ferroelectrics and ferroelastics. III. Application to $Gd_2(MoO_4)_3$. *J. Phys. Soc. Japan.* **56**, 596.

SYOYAMA, S. & OSAKI, K. (1972). X-ray study of low temperature form of $MgSiF_66H_2O$ and the relation between crystal lattice of low and high temperature forms.*Acta Cryst.B* **28** 2626

SVENSSON, C. & ABRAHAMS, S.C. (1984). Phase transition in, and ferroelasticity of 9-hydroxyphenalenone. *Phase Transformations in Solids Symp. Proc.* , 149.

TANAKA, M. & TATSWZALIS, I. (1983). Raman scattering studies of the phase transition in $KH_3(SeO_3)_2$ and $KD_3(SeO_3)_2$. *Ferroelectrics* **52(1-3)**, 195.

TELLO, M.J., MANES, J.C., FERNANDEZ, J., ARRIANDIAGA, M.A. & PEREZ-MATO, J.M. (1981). Physical properties and phenomenological theory around the successive high-temperature phase transitions of $(C_2H_5NH_3)_2CuCl_4$. *J. Phys. C* **14(6)**, 805.

TING, C. & GUANG-YAN, H. (1987). Raman spectroscopic study of ferroelastic phase transition in lanthanide pentaphosphates. *Chin. Phys.* **7(2)**, 422.

TING, C. & GUANG-YAN, H. (1986). Raman spectroscopic study of ferroelastic phase transition in lanthanide pentaphosphates. *Acta Phys. Sin.* **35**, 1521.(In Chinese).

TOLEDANO,J .C., ERRANDONEA, G., & JAGUIN J.P.(1976). Soft acoustic mode in ferroelastic lanthanum pentaposphate.*Sol.State Comm..***20**,905.

TOLEDANO, P., FEJER, M.M. & AULD, B.A. (1983). Nonlinear elasticity in proper ferroelastics. *Phys. Rev. B* **27(9)**, 5717.

TORRES, J. & JOFFRIN, C. (1980). Inelastic neutron scattering study of the ferroelastic transition in lead phosphovanadate $Pb_3(P_{0.95}V_{0.05}O_4)_2$. *Ferroelectrics* **26(1-4)**, 665.

TORRES, J., ROUCAU, C. & AYROLES, R. (1982). Investigation of the interactions between ferroelastic domain walls and of the structural transition in lead phosphate observed by electron microscopy.I. Experimental results. *Phys. Status Solidi a* **70(2)**, 659.

TORRES, J., ROUCAU, C. & AYROLES, R. (1981). Electron microscope observations of the interactions between ferroelastic domain walls and the phase transition in lead phosphate. *Symmetry and Broken Symmetry in Condensed Matter Physics (Proc. Coll. Pierre 1980) Curie, IDSET* , 323 (In French).

TORRES, J., PRIMOT, J., POUGNET, A.M. & AUBREE, J. (1980a). Differential thermal analysis, dilatometric and Brillouin scattering measurements at the ferroelastic phase transition of lead phosphovanadate compounds. *Ferroelectrics* **26(1-4)**, 689.

TORRES, J., AYROLES, R., ROUCAU, J. & TALIANA, M. (1980b). Electron microscopy study of the domain structure in the ferroelastic phase of lead phosphate $Pb_3(PO_4)_2$. *Ferroelectrics* **29(1-2)**, 63.

TUSZYNSKI, J.A., MROZ, B., KIEFTE, H. & CLOUTER, M.T. (1988). On the ferroelastic phase transition in $LiCsSO_4$. II. Theoretical model. *Ferroelectrics* **81**, 1183.

UDAGAWA, M., KOHN, K. & NAKAMURA, T (1978). Brillouin scattering study on lithium ammonium-tartrate monohydrate in paraelectric phase.*J.Phys.Soc Jap.* **44** 1873.

UNRUH, H.G., MÜHLENBERG, D. & HAHN, Ch. (1991). Ferroelastic phase transition in $CaCl_2$ studied by Raman spectroscopy. *J. Phys. Cond. Matt.,* in press

VACHER, R., BOISSIER, M. & SAPRIEL, J. (1981). Brillouin-scattering investigation of the ferroelastic transition of benzil. *Phys. Rev. B* **23(1)**, 215.

VAGIN, S.V., DUDNIK, E.F. & SINYAKOV, E.V. (1979). Statics and dynamics of domains in the ferroelastic $Pb_3(PO_4)_2$. *Bull.Acad.Sci.USSR, Phys. Ser.* **43(8)**, 160.

VERWERFT, M., BRODDIN, D., VAN TENDELOO, G., VAN LANDUYT, J. & AMELINCKX, S. (1989). Electron microscopy of phase transitions. *Phase Transitions* **14(1-4)**, 285.

VLOKH, O.G., KAMINSKII, B.V., POLOVINKO, I.I., SVELEVA, S.A., KHALAKHAN, A.YU., BOGDANOVA, A.V. & PETROV, V.V. (1988). Phase diagrams and piezo-optic effect in Cs_2HgBr_4 crystals. *Sov. Phys.* **30(6)**, 1102.

VON HODDENBERG, R. & SALJE, E. (1977). Temperature dependence of critical stress and phase diagram of ferroelastic $Pb_3(PO_4)_2$-$Pb_3(VO_4)_2$. *Mat. Res. Bull.* **12**, 1029.

VON DER MUHLL, R., RAVEZ, J., HAGENMULLER, P., BARBIER, P., DRACHE, M. & MAIRESSE, G. (1982). Cs_3BiCl_6: a ferroelastic phase having a remanent polarization at ambient temperatures. *Solid State Commun.* **43(11)**, 797 (In French).

WADWAHAN, V.K. (1980). Is potassium chlorate a ferroelastic? *Acta Crystallogr.* **A36(6)**, 851.

WADWAHAN, V.K. & GLAZER, A.M. (1989). Prototype symmetry of the ferroelastic superconductor Y-Ba-Cu-O. *Phys. Rev. B* **39**, 9631.

WADWAHAN, V.K. & SOMAJAZALU, M.S. (1986). Symmetry analysis of the atomic mechanism of ferroelastic switching and mechanical twinning. Application to thallous nitrate. *Phase Transitions* **7**, 59.

WAINER, L.S., BAGGIO, R.F., DUSSEL, H.L. & BENYACA, M.A.R. (1981). Study of domains and domain walls in ferroelastic $BiVO_4$. *Ferroelectrics* **31**(3-4), 121.

WENYUAN, S., HUMIN, S., YENING, BAOSHENG, L. (1985). Internal friction associated with domain walls and ferroelastic phase transition in LNPP.*J. Phys. Colloq.* **46**, 609.

WOLEJUO, T., PAKULSKI, G. & TYLIZYNSKI, Z. (1988). Ferroelastic properties of $LiRb_5(SO_4)_3.1.5H_2SO_4$ crystal. *Ferroelectrics* **81**, 1143.

WONG, P.T.T. (1980). Molecular distortion and ferroelasticity in $K_2Hg(CN)_4$ investigated by Raman Scattering. *Solid State Commun.* **36**(2), 185.

WOOD, I.G., WADWAHAWAN, V.J. & GLAZER, A.M. (1980). Temperature dependence of spontaneous birefringence in ferroelastic lead orthophosphate. *J. Phys. C* **13**(27), 5155.

XUXIN, Z., JIANXIANG, S., WEIJI, J. (1986). Detection of ferroelastic phase transition of $LiKSO_4$ crystal by depolarization of transmitted light. *Chin. Phys. Lett.* **3**, 325.

YACOUBI, A., RAVEZ, J. & GRANNEC, J. (1987). $Na_5Ti_3O_3Fl_1$: a new ferroelastic oxylfluoride.*Chem.Scr.* **27**(3),429.

YA-GU, W.&YE-MING,W. (1985). Ultrasonic attenuation and internal friction during ferroelastic transformation of $Gd_2(MoO_4)_3$ crystals. *Acta Phys. Sin.* **34**, 520. (In Chinese) (Translation in *Chin. J. Phys.* 34).

YAMADA, Y., NODA, Y. & ENDON, Y. (1983). Proton dynamics at the ferroelastic phase transition in $KH_3(SeO_3)_2$. *Physica B&C* **120B+C**, 270.

YENING, W., WENYUAN, S., XIACHUA, C., HUIMIS, S. & L. BAOSHING. (1987). Internal friction associated with the domain walls and the second-order ferroelastic transition in CNPP. *Phys. Status Solidi A* **102**(1), 279.

YOKOTA, S. (1982). Ferroelastic phase transition of $CsHSeO_4$. *J. Phys. Soc. Japan*. **51**(6), 1884.

YOKOTA, S. & MAKITA, Y. (1982). Ferroelasticity in $CsHSeO_4$ below the 128°C transition. *J. Phys. Soc. Japan.* **51**(1), 138.

YOKOTA, S., MAKITA, Y. & TAKAGI, Y. (1982). Ferroelastic phase transition in $K_3H(SeO_4)_2$. *J. Phys. Soc. Japan.* **51**(5), 1461.

YOSHIDA, H., TAKEMASA, H. OSHINO, Y. & MAKITA, Y. (1984a). Calorimetric and dilatometric studies on successive phase transitions in ferroelastic NH_4BeF_3. *J. Phys. Soc. Japan.* **53**(8), 2600.

YOSHIDA, H., ENDO, M., KANEKO, T., OSAKA, T. & MAKITA, Y. (1984b). Ferroelastic phase transition in TlH_2PO_4. *J. Phys. Soc. Japan.* **53**(3), 910.

YOSHIHARA, A., YOSHIZAWA, M., YASUDA, H. & FUJIMARA, T. (1985). On the 83K ferroelastic phase transition in benzil. *Japan. J. Appl. Phys. Suppl.* **24**, 367

ZAPART, M.B. & ZAPART, W. (1988). Uniaxial stress effect on ferroelastic phase transition in KSc(MoO$_4$)$_2$. *Ferroelectrics* **80**, 711.

ZAPART, W. & ZAPART, M.B. (1988). EPR investigation of the pseudoproper ferroelastic phase transition in KH$_3$(SeO$_3$)$_2$. *Ferroelectrics* **80**, 707.

ZAPART, M.B., ZAPART, W., STANKOWSKI, J. & ZVIAGIN, A.I. (1982). Ferroelastic phase transitions in KSc(MoO$_4$)$_2$ monocrystals by electron paramagnetic resonance of Cr^{3+}-ions. *Physica B&C* **114B+C(2)**, 201.

ZEKS, B., LAWRENCIC, B.B. & BLINC, R. (1984). Microscopic model for the ferroelastic transition in KLiSO$_4$. *Phys. Status Solidi B* **122(2)**, 399.

ZUNIGA, F.J., TELLO, M.J., PEREZ-MATO, J.M., PEREZ-JUBINDO, M.A. & CHAPUIS, G. (1982). Second order ferro-paraelastic phase transition in the layer crystal (n-C$_3$H$_7$NH$_3$)$_2$ZnCl$_4$ at 310K. *J. Chem. Phys.* **76(5)**, 2610.

INDEX

acoustic instability, 112ff
adularia, 117
Aizu strain, 27
Albite twin law, 51, 54, 56ff, 114
amplitudon, 169
anorthite, 30, 39, 40, 47, 72, 104, 121, 143, 148
antiferrodistortive, 44
anti-phase boundaries, 75
A_2O_5, 21, 37, 38, 41, 47
$AgNa(NO_2)_2$, 120
average structure (strain), 24

$Ba_2NaNb_5O_{15}$, 90
$BaMnF_3$, 132
bilinear coupling, 29, 30ff, 141, 151ff, 156, 176, 188, 210
biquadratic coupling etc., 29, 144, 151ff, 161ff, 176
$BiVO_4$, 34
birefringence, 6, 41, 138

Cahn equation, 208ff
calcite, 10, 11, 53, 104, 179
Ca-feldspar, 143, see also anorthite
$Cd_2(NH_4)_2(SO_4)_3$, 53
$Cd_2Tl_2(SO_4)_3$, 53
co-elastic, 9, 12
coupling, 28, 29, 30, 46, 141, 151, 156, 164, 176, 187, 188, 202, 210
coupling theory, 28ff
crest-riding periodon, 200
critical radius, 89
critical stress, 5
coercive stress, 5, 9
cordierite, 192
cyanides, 35

defect interactions, 125, 133ff
defect segregation, 134ff
devil's staircase, 174
dislocation density, 55ff
domain mobility, 76ff
domain switching, 3
dressed fluctuation, 194

effective Hamiltonian, 79, 123, 126
elastic constants, 29ff, 33, 42, 114
elastic hysteresis, 5
embryo, 193

entropy, 14ff, 118ff

feldspars, 29ff, 39ff, 47, 52ff, 67, 97, 104 121, 141, 143, 148, 172, 202, 207
Fe_3O_4, 132
ferrobielastic, 46
ferrodistortive, 44
ferroelastic - definition, 5, 46
ferroelastic switching, 3
ferroelastoelectrics, 46
ferromagnetoelastic, 46
ferroic phase transition, 44
Frenkel-Kontorova model, 174
fluctuations (and specific heat), 122ff

$Gd_2(MoO_4)_3$, 60
gradient coupling, 164ff, 187
Ginzburg energy, 93, 164ff, 187ff,

hard ferroelastics, 10
H_3BO_3, 53
Heine-McConnell model, 186ff
herringbone patern, 167
High-T_c superconductors, 65, 73, 78, 85, 86, 109
hyperplastic, 12
hysteresis - elastic, 3
hysteresis - optical, 3

In-Tl, 90
incommensurate phase, 164ff, 182
intersection of domain walls, 59ff, 109ff

KH_2PO_4, 36, 104, 132
kinks, 84ff
kinetic order parameter, 203
kinetic rate laws, 87, 194, 202ff, 205, 208, 217

$LaNbO_4$, 34
Lagrange equation, 94
Landau-Khalatnikov equation, 80
Landau potential, 13ff, 119ff
langbeinites, 122, 138ff
lattice imperfections, 125, 127, 133
lead arsenate, 128,
lead phosphate, 5, 43, 47, 73, 113, 134, 136, 139, 169, 171, 193
lead phosphate-vanadate, 6, 104
lead vanadate, 67, 77, 83, 102

Index

Printed in the United States
By Bookmasters